MAPPING WORKBOOK

Karl Byrand

Green River Community College

GLOBALIZATION AND DIVERSITY

GEOGRAPHY OF A CHANGING WORLD

THIRD EDITION

Rowntree • Lewis • Price • Wyckoff

Prentice Hall

Boston Columbus Indianapolis New York San Francisco Upper Saddle River
Amsterdam Cape Town Dubai London Madrid Milan Munich Paris Montréal Toronto
Delhi Mexico City São Paulo Sydney Hong Kong Seoul Singapore Taipei Tokyo

Acquisitions Editor: Christian Botting
Editor in Chief, Chemistry and Geosciences: Nicole Folchetti
Marketing Manager: Maureen McLaughlin
Assistant Editor: Jennifer Aranda
Managing Editor, Chemistry and Geosciences: Gina M. Cheselka
Project Manager: Traci Douglas
Operations Specialist: Maura Zaldivar
Supplement Cover Designer: Paul Gourhan
Cover Illustrator: Quade Paul

Pearson Prentice Hall
Upper Saddle River, NJ 07458

Printed in the United States of America

10 9 8 7 6 5 4 3 2 1

ISBN-13: 978-0-321-66739-7
ISBN-10: 0-321-66739-5

Prentice Hall
is an imprint of

www.pearsonhighered.com

Table of Contents

Chapter One: Diversity Amid Globalization Mapping Workbook Exercises 1

Chapter Two: The Changing Global Environment Mapping Workbook Exercises 15

Chapter Three: North America Mapping Workbook Exercises 23

Chapter Four: Latin America Mapping Workbook Exercises 39

Chapter Five: Caribbean Mapping Workbook Exercises 53

Chapter Six: Sub-Saharan African Mapping Workbook Exercises 69

Chapter Seven: Southwest Asia and North Africa Mapping Workbook Exercises 89

Chapter Eight: Europe Mapping Workbook Exercise 105

Chapter Nine: The Russian Domain Mapping Workbook Exercises...................... 119

Chapter Ten: Central Asia Mapping Workbook Exercises................................. 135

Chapter Eleven: East Asia Mapping Workbook Exercises..................................153

Chapter Twelve: South Asia Mapping Workbook Exercises 169

Chapter Thirteen: Southeast Asia Mapping Workbook Exercises 183

Chapter Fourteen: Australia and Oceania Mapping Workbook Exercises 201

Chapter One: Diversity Amid Globalization Mapping Workbook Exercise

Identify the following world regions on workbook Map 1.1

Australia and Oceania
Central Asia
East Asia
Europe
North Africa/Southwest Asia
North America
South America
South Asia
Southeast Asia
Sub-Saharan Africa
The Caribbean
The Russian Domain

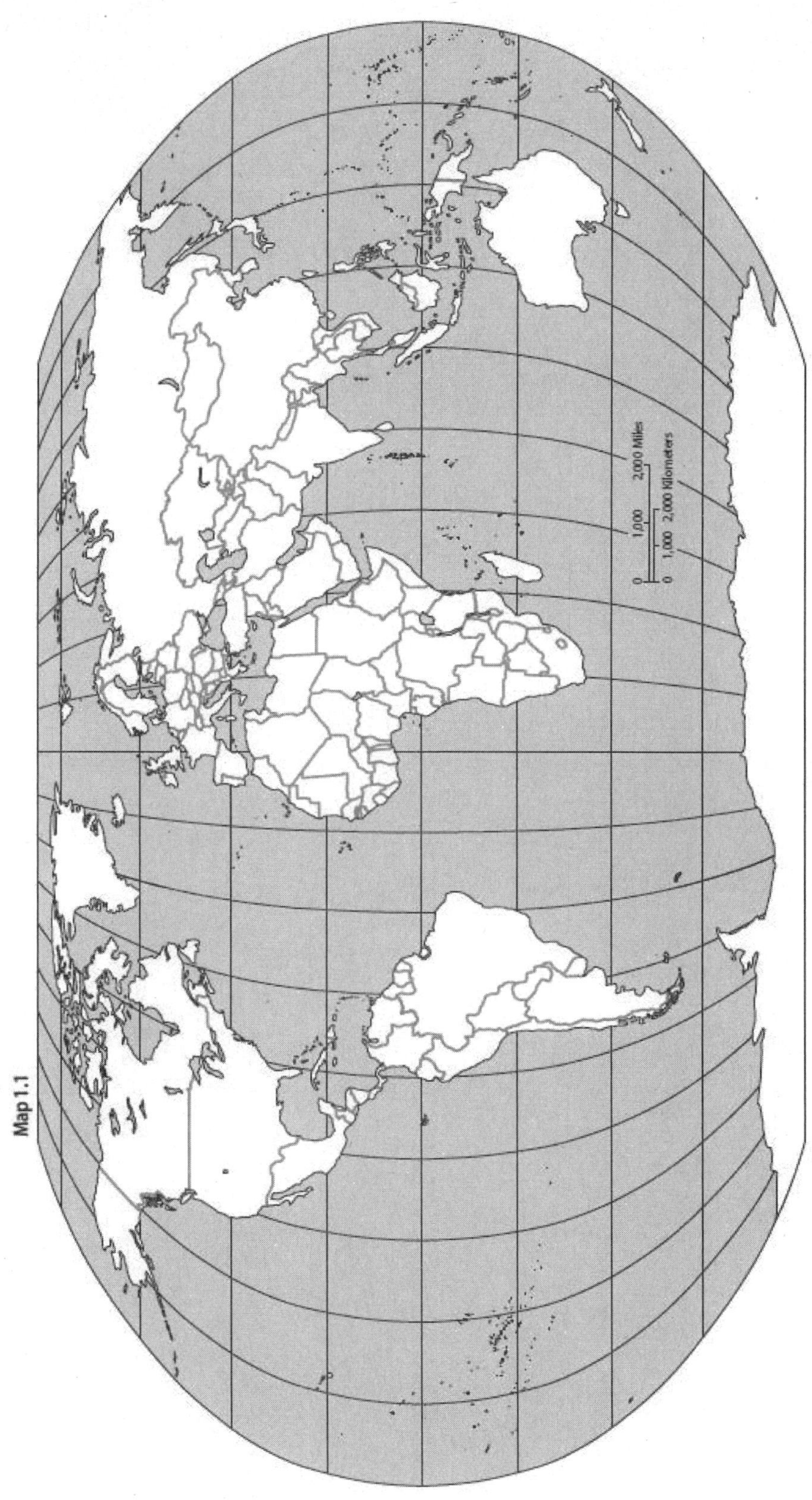

Map 1.1

Exercise One: Population Indicator Comparison

Using Table 1.1 Population Indicators of the 10 Largest Countries (p. 12) and Mapping Workbook Map 1.2, complete the following exercise.

Using the data provided in the Population Density (per square kilometer density) column of Table 1.1 "Population Indicators of the 10 Largest Countries" (p. 12), use a colored pencil to shade in the countries with a per kilometer density of 8.0–77.6. Using different colored pencils for each category, shade the countries that possess a per kilometer population density of 77.7–147.2, 147.3–216.8, 216.9–286.4, and 286.5 and above. Mark your shading scheme in the map's legend. Once you have done this, answer the following questions.

1. Of the mapped countries, which have the highest population density?

____

2. Is there a regional pattern associated with countries possessing high population densities?

____

3. Of the mapped countries, which have the lowest population density?

____

4. Does there appear to be a regional pattern associated with countries possessing low population densities?

____

Compare those countries you shaded with the population distributions displayed in Figure 1.16 "World Population" (p. 12-13).

5. Would you say that the population is evenly distributed (i.e., dispersed) within the majority of countries, or does it appear to display a concentration?

____

6. Explain the reasons for either observed pattern. More specifically, is there one or more common landscape characteristics that tend to determine where population is concentrated?

____

____

____

____

____

____

Using the data provided in the total fertility rate (TFR) column of Table 1.1 "Population Indicators of the 10 Largest Countries" (p. 12), use a colored pencil to pattern in the countries with a TFR of 1.4–2.26. Using different colored pencils for each category, pattern the countries that possess TFRs of 2.27–3.12, 3.13–3.98, 3.99–4.84, 4.85–5.7. Mark your patterning scheme in the map's legend. Once you have done this, answer the following questions.

7. Compare your population density map with the map of total fertility rates you created. Does there appear to be a correlation between countries possessing high total fertility rates and high population densities? Please specify which countries display this pattern.

8. Does there appear to be a correlation between countries possessing low total fertility rates and low per square kilometer population densities? Please specify which countries display this pattern.

9. Are there outliers to either of these scenarios (i.e., countries with high population densities and low total fertility rates, and vice versa)? If so, name these countries and explain the possible reasons as to why this is the situation.

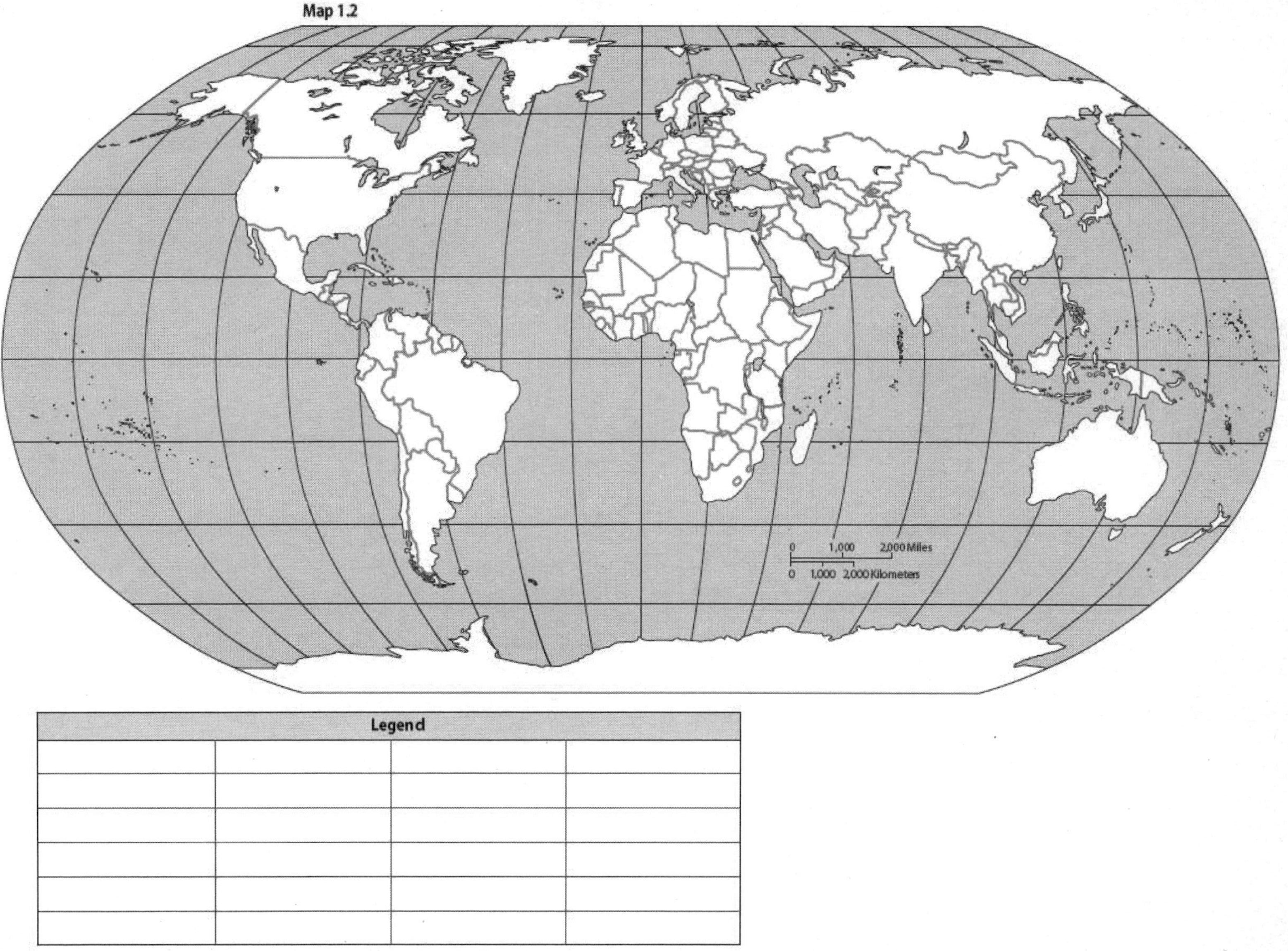
Map 1.2
0 1,000 2,000 Miles
0 1,000 2,000 Kilometers
Legend

Exercise Two: Development Indicator Comparison

Using Table 1.2 Development Indicators of the 10 Largest Countries (p. 30) and Mapping Workbook Map 1.3, complete the following exercise.

Using the data provided in the GNI Per Capita, PPP 2007 column of Table 1.2 "Development Indicators of the 10 Largest Countries" (p. 30), use a colored pencil to shade in the countries with a GNI of 1,000–2,000. Using different colored pencils for each category, shade the countries that possess GNIs of 2,001–4,000, 4,001–6,000, 6,001–10,000, and 10,001 and above. Mark your shading scheme in the map's legend. Once you have done this, answer the following questions.

Using the data provided in the Human Development Index (HDI), 2006 column of Table 1.2 "Development Indicators of the 10 Largest Countries" (p. 30), use a colored pencil to shade in the countries with a HDI of 0.499–0.5904. Using different colored pencils for each category, shade the countries that possess HDIs of 0.5905–0.6818, 0.6819–0.7732, 0.7733–0.8646, 0.8647–0.956. Mark your shading scheme in the map's legend. Once you have done this, answer the following questions.

1. Define GNI.

____

____

2. Define HDI.

____

____

3. Is one of these development indicators a better measure than the other? If so, which one and why?

____

____

____

4. Among the ten most populous countries, is there a regional pattern of high, medium, and low GNI countries? If so, what is that pattern?

____

____

____

5. Among the ten most populous countries, is there a regional pattern of high, medium, and low HDI countries? If so, what is that pattern?

____

____

____

6. Now observe the maps together. What correlation is there between countries whose development is measured by GNI as compared to HDI? For example, do the same countries that possess a high HDI also display a high GNI and countries that display a low HDI display a low GNI?

____

____

____

7. Name the countries that display this correlation.

8. Looking at the maps together once again, do you notice outliers to the above? Specifically, are there countries that have a different ranking with their GNIs and the HDIs?

9. Name any countries that display this pattern.

10. What do you think would account for such differences in GNI and HDI in outlier countries?

11. Based on this, which do you believe is a better measure of development? Why?

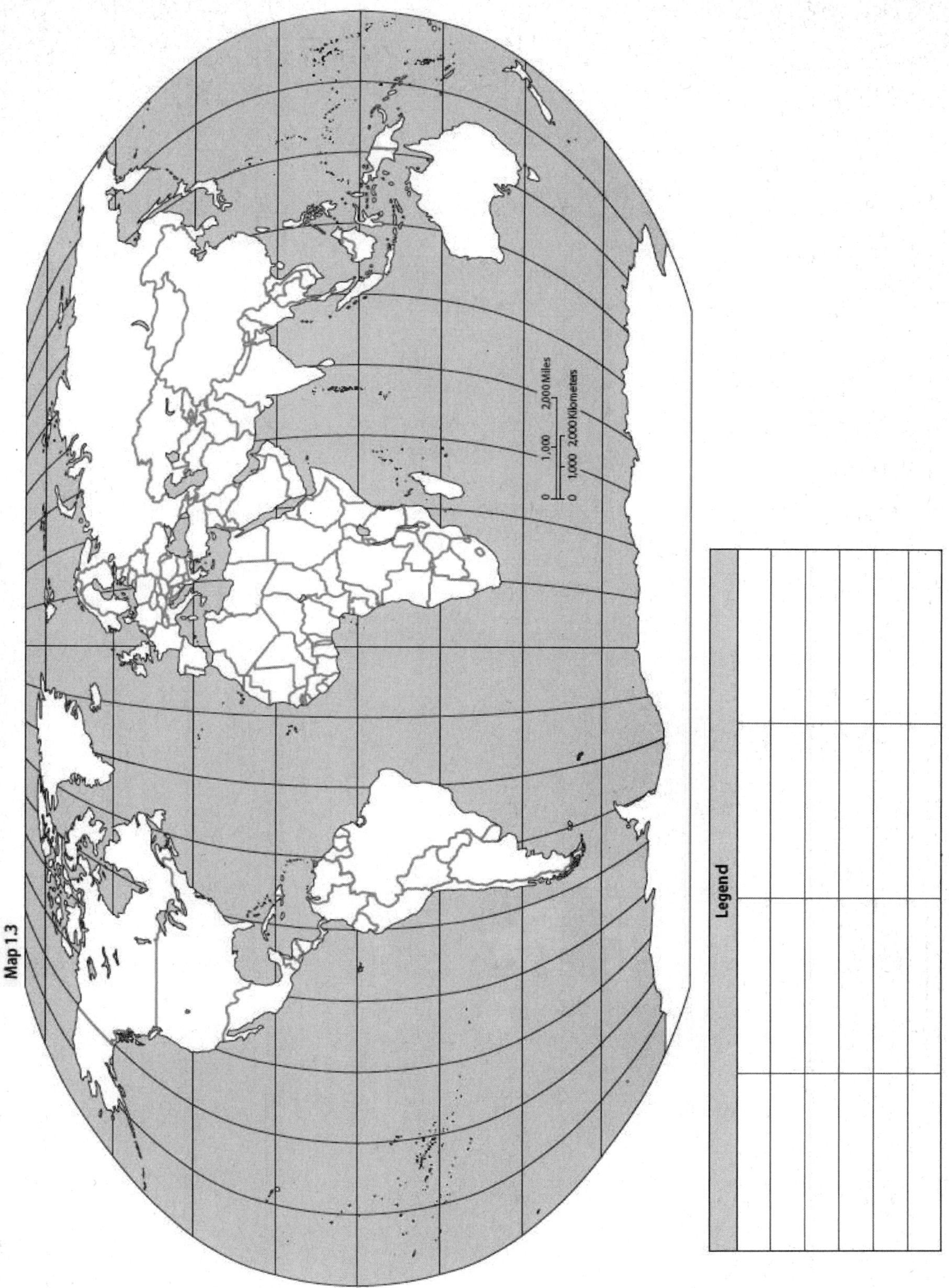
Map 1.3
0 1,000 2,000 Miles
0 1,000 2,000 Kilometers
Legend

Bonus Exercise: Create a third map based on the development indicator of your choosing from Table 1.2 and compare it against the GNI and HDI of the ten most populace countries. Discuss the correlations, the outliers, and the reasons why they occur.

Table

GNI	Color	HDI	Color
1,000–2,000		0.499–0.5904	
2,001–4,000		0.5905–0.6818	
4,001–6,000		0.6819–0.7732	
6,001–10,000		0.7733–0.8646	
10,001 and above		0.8647–0.956	

Chapter Two: The Changing Global Environment Mapping Workbook Exercises

Exercise One: Regional Impacts of Global Warming

Using Table 2.1 The World's Major CO_2 Polluters (p. 41) and Mapping Workbook Map 2.1, complete the following exercise.

Using the data provided in the Per Capita Emissions of CO_2, Metric Tons column of Table 2.1 "The World's Major CO_2 Polluters" (p. 41), use a colored pencil to shade in the countries with a per capita emission of CO_2 of 1.2–4.65 metric tons. Using different colored pencils for each category, shade the countries that possess a per capita emission of CO_2 of 4.66–9.3 metric tons, 9.4–13.95 metric tons, and 13.96–19.8 metric tons. Mark your shading scheme in the map's legend. Once you have done this, answer the following questions.

1. Is there a geographic pattern associated with these polluting countries? If so, what is it?

___

___

___

2. Do these nations possess the same level of economic and industrial development?

___

___

___

Now read the section called "Human Impacts on Plants and Animals: The Globalization of Nature" (p. 41).

On your map use a red or black (or another color that will contrast with your above shading scheme) colored pencil to pattern in the countries/regions experiencing the highest levels of tropical deforestation. Mark your shading scheme in the map's legend.

Now read the section called "Deserts and Grasslands" (p. 44).

On your map use a red or black (or another color that will contrast with your above shading and patterning schemes) colored pencil to pattern in the countries/regions experiencing the highest levels of desertification. Mark your shading scheme in the map's legend. Once you have done this, answer the following questions.

3. What are the primary causes of tropical deforestation?

___

___

___

4. What are the primary causes of desertification?

___

___

___

Compare the areas that are experiencing tropical deforestation and desertification with the pattern of the major CO_2 producers.

5. Is there a correlation between these two patterns (e.g., are these producers the same nations that are experiencing the highest rates of tropical deforestation and desertification)?

Another significant problem associated with global warming is sea level rise. On your map use a red or black (or another color that will contrast with your above shading and patterning schemes) colored pencil to pattern the coastal areas of those countries that are the major CO_2 producers. Mark your shading scheme in the map's legend.

Now examine Figure 1.16 World Population (pp. 12-13). Once you have done this, answer the following questions.

6. Is there a correlation between the regions of highest population density and coastal areas in the major CO_2 producing countries? If so, what is it?

7. It is estimated that sea levels could rise 1 meter (3.3 feet) by the year 2100. If this holds true, how will sea level rise have an impact on these CO_2 producing nations?

8. Will there also be an impact on the nations that produce only small quantities of CO_2?

9. How equitable is this situation?

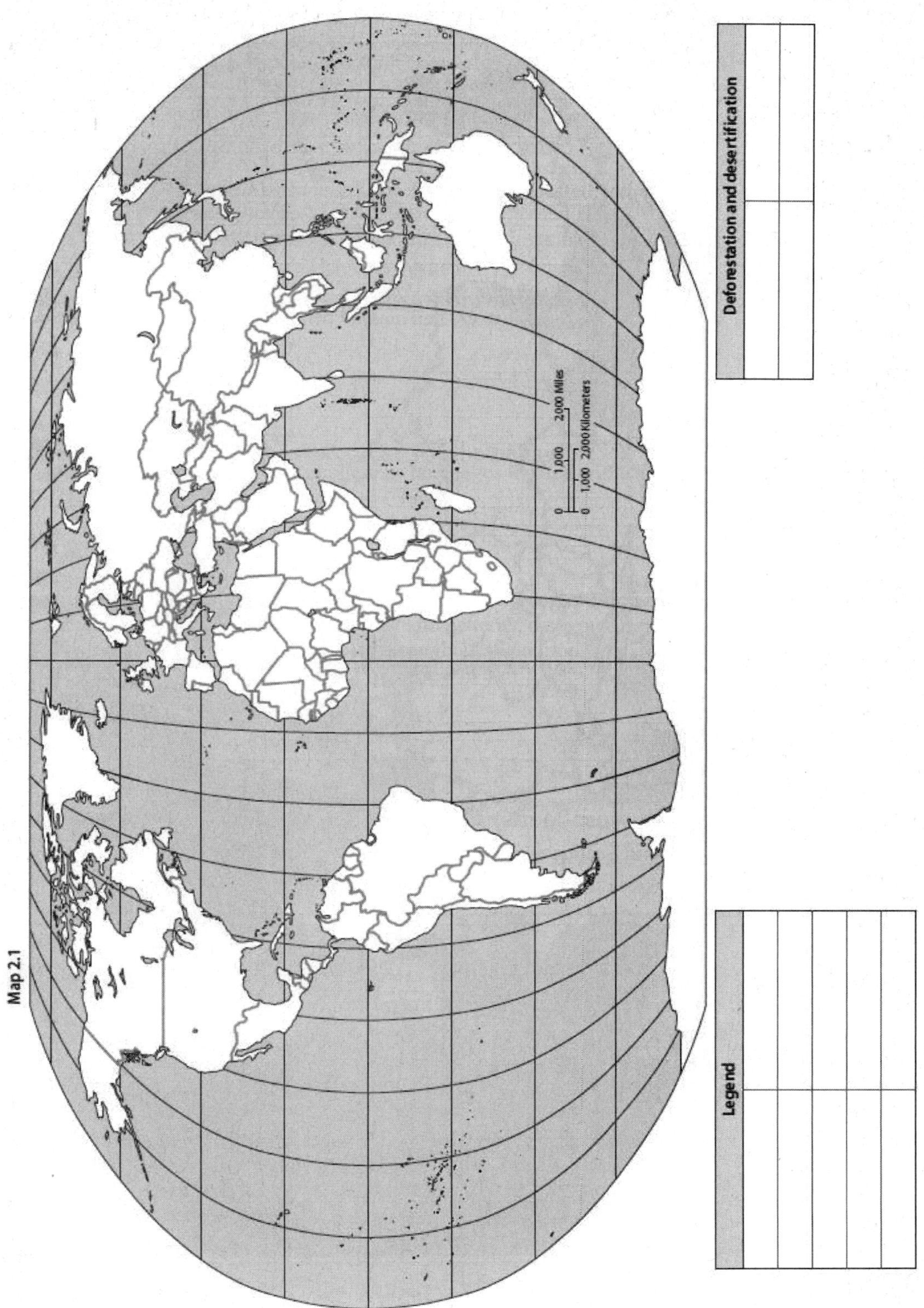
Map 2.1
0 1,000 2,000 Miles
0 1,000 2,000 Kilometers
Deforestation and desertification
Legend

Exercise Two: Climate as a Biome Control

Using Figure 2.3 World Climate Regions (pp. 38-39), Figure 2.8 World Bioregions (pp. 42-43), Figure 1.16 World Population (pp. 12-13), and Mapping Workbook Map 2.2, complete the following exercise.

Using Figure 2.3 "World Climate Regions" (pp. 38-39) as a reference, use different colored pencils to shade in the approximate locations of the major climatic regions. Mark your shading pattern in the map's legend. Using Figure 2.8 "World Bioregions" (pp. 42-43) as a reference, use different colors to represent each bioregion, pattern in the various bioregions in their respective locations. Mark your patterning scheme in the map's legend. Once you have done this, answer the following questions.

In the table below, specifically match each bioregion with its mapped climatic region.

Bioregion	Climate Region

1. What is the general pattern of world climatic regions?

____

____

____

2. What are the general controls (i.e., causes or determinants) of these climate regions?

____

____

____

3. What is the general correlation between the climatic regions and the world bioregions?

____

____

____

4. What characteristics of these climatic regions determine these bioregions?

Using Figure 1.16 "World Population" (pp. 12-13) as a reference, use a red or black (or another color that will contrast with your above shading and patterning scheme) colored pencil to shade in the regions possessing the highest population density.

5. In which climate region(s) do the majority of the planet's population reside?

6. What are the characteristics of this (these) climatic region(s)?

7. Why do you think the majority of the planet's population resides within this (these) regions(s)?

Examine the regions of low population density depicted on Figure 1.16 "World Population" (pp. 12-13).

8. In which climate region(s) do we see these lower population densities?

9. What are the characteristics of this (these) climatic region(s)?

10. Why do you think this (these) climatic region(s) possess lower population densities?

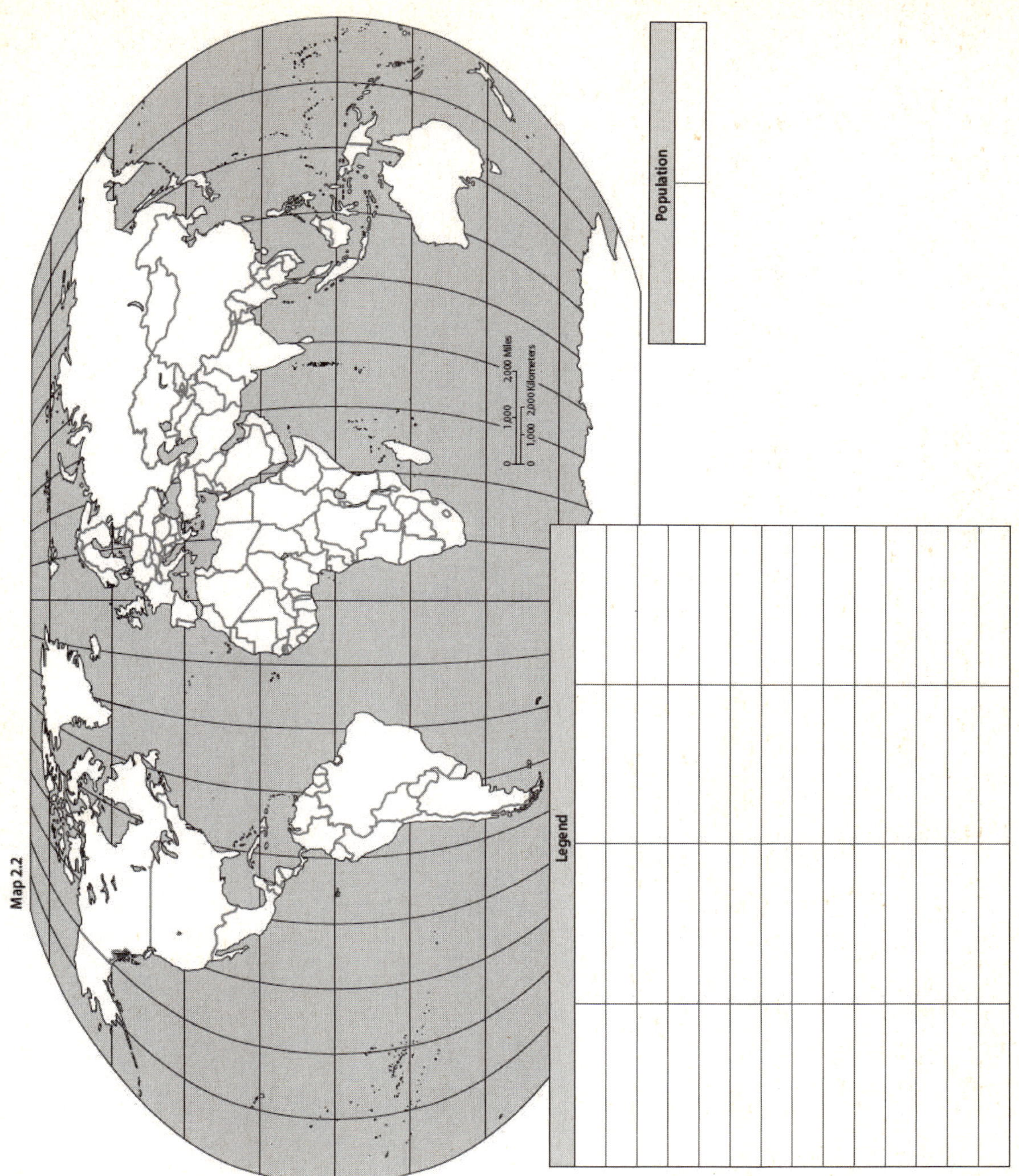
Map 2.2
Population
0 1,000 2,000 Miles
0 1,000 2,000 Kilometers
Legend

Chapter Three: North America Mapping Workbook Exercises

Identify the following features on workbook Maps 3.1 and 3.2

Identify and label the following countries on Map 3.1

Canada
United States

Identify and label the following states and provinces on Map 3.1

Alabama
Alaska
Alberta
Arizona
Arkansas
British Columbia
California
Colorado
Connecticut
Delaware
Florida
Georgia
Hawaii
Idaho
Illinois
Indiana
Iowa
Kansas
Kentucky
Louisiana
Maine
Manitoba
Maryland
Massachusetts
Michigan
Minnesota
Mississippi
Missouri
Montana
Nebraska
Nevada
New Brunswick
New Hampshire
New Jersey
New Mexico
New York
Newfoundland and Labrador
North Carolina
North Dakota
Northwest Territories
Nova Scotia
Nunavut
Ohio
Oklahoma
Ontario
Oregon
Pennsylvania
Prince Edward Island (P.E.I.)
Quebec
Rhode Island
Saskatchewan
South Carolina
South Dakota
Tennessee
Texas
Utah
Vermont
Virginia
Washington
Washington, D.C.
West Virginia
Wisconsin
Wyoming
Yukon Territory

Identify and label the following cities on Map 3.2

Albuquerque
Anchorage
Atlanta
Boston
Calgary
Charlotte
Chicago
Columbus
Dallas-Fort Worth
Denver
Detroit
Halifax
Honolulu
Houston
Indianapolis
Kansas City
Las Vegas
Los Angeles
Louisville
Memphis
Miami
Minneapolis-St. Paul
Montreal
Nashville
New Orleans
New York City
Oklahoma City
Ottawa
Philadelphia
Phoenix
Portland
Quebec City
Regina
Salt Lake City
San Diego
San Francisco
Seattle
Saint John
St. Louis
Tampa
Toronto
Whitehorse
Winnipeg
Vancouver

Identify and label the following physical features on Map 3.2

Alaska Range
Appalachian Highlands
Arctic Ocean
Atlantic Ocean
Baffin Bay
Baffin Island
Bering Sea
Brooks Range
Cascade Range
Central Valley
Chesapeake Bay
Coast Mountains
Coast Ranges
Colorado Plateau
Columbia River
Florida Keys
Great Plains
Gulf of Alaska
Gulf of Mexico
Hawaii
Hudson Bay
Kauai
Lake Erie
Lake Huron
Lake Michigan
Lake Ontario
Lake Superior
Mackenzie River
Maui
Mississippi River
Missouri River
Newfoundland
Oahu
Ohio River
Ozark Mtns.
Pacific Ocean
Prudhoe Bay
Rio Grande River
Rocky Mountains
Sierra Nevada
St. Lawrence River
Teton Range
Vancouver Island

0 250 500 Miles
0 250 500 Kilometers
0 100 Miles
0 100 Kilometers
155°W
160°W
20°N

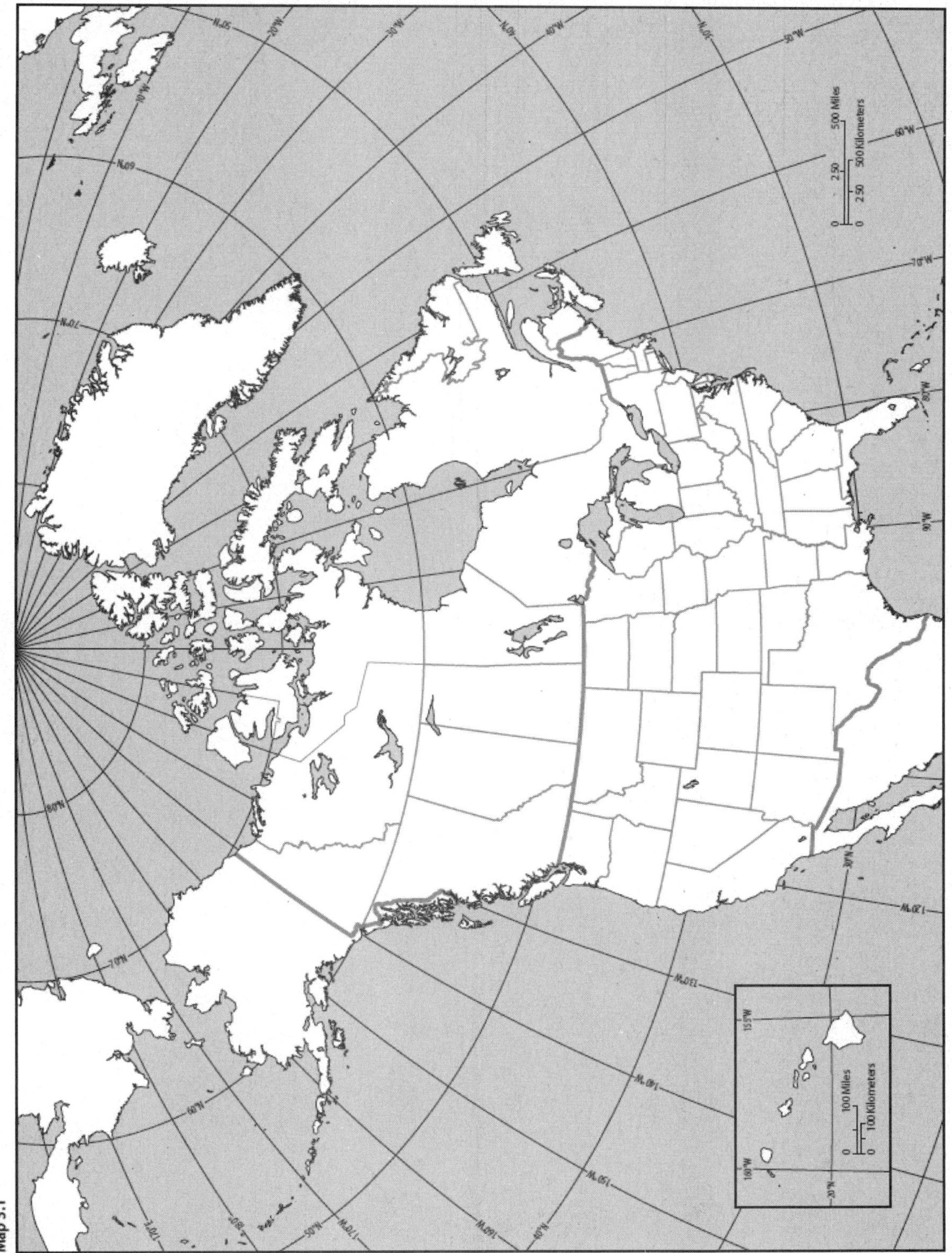

Map 3.1

0
250
500 Miles
0
250
500 Kilometers
0
100 Miles
0
100 Kilometers

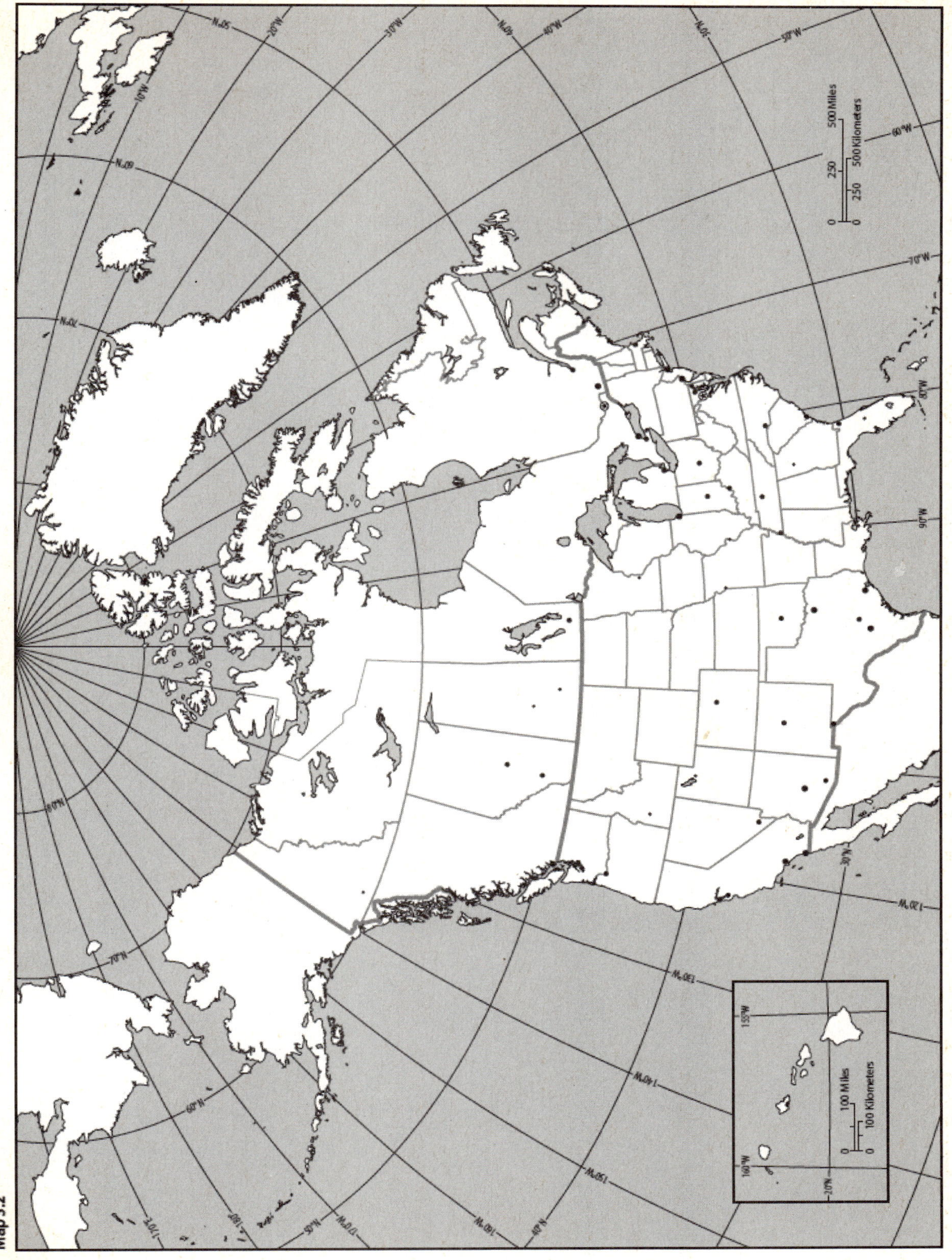

Map 3.2

Exercise One: Internal Patterns of North American Migration

Using Figure 3.1 North America (p. 51), Figure 3.10 European Settlement Expansion (p. 60) and Mapping Workbook Map 3.3, complete the following exercise.

Read "Occupying the Land" (pp. 58-60) in your textbook. Using Figure 3.10 "European Settlement Expansion" (p. 60) as a reference, create an isochronic (line of equal time) map. Begin by using a colored pencil and draw a line indicating the North American regions that were settled by 1750. Label this line "1750." Using a different colored pencil, draw a line indicating the North American regions that were settled between 1750 and 1850. Label this line "1750-1850." Using a different colored pencil, draw a line indicating the North American regions that were settled by 1910. Label this line "1910." Mark the lines in the map's legend. Once you have done this, answer the following questions.

1. Prior to 1750, what were the major North American regions that were settled?

____

2. What were the factors explaining this settlement pattern?

____

____

3. What were the major North American regions settled between 1750 and 1850?

____

4. What were the factors explaining this settlement pattern?

____

____

5. What were the major North American regions settled between 1850 and 1910?

____

6. What were the factors explaining this settlement pattern?

____

____

Examine Figure 3.1 "North America" (p. 51), paying close attention to the continent's physical geography.

7. How do you think North America's physical landscape has had an impact on the above settlement patterns?

____

____

8. At what point in time, if any, do you believe that the physical landscape would have been less of a determinant associated with this pattern of westward settlement?

____

____

9. If you believe that it eventually became less of a determinant, explain the reasons behind its decreased impact.

____

____

Read "North Americans on the Move" (pp. 60-62) in your text. Using a colored pencil, create flow arrows indicating the movement of Blacks from the rural South to the urban North. Label the arrow with the approximate period that this migration occurred. Mark this symbol in the map's legend.

Using a different colored pencil, create a flow arrow indicating the general pattern of regional growth in the Sun Belt South. Label the arrow with the approximate period that this migration occurred. Mark this symbol in the map's legend.

Using a different colored pencil, create a flow arrow indicating the general pattern of regional growth in the mountain states. Label the arrow with the approximate period that this migration occurred. Mark this symbol in the map's legend. Once you have done this, answer the below questions.

10. What was the major period when Blacks migrated from the rural South to the urban North?

____

11. What were the contributing factors associated with this migration?

____

____

12. What were the impacts of this migration?

____

____

13. What was the major period when the Sun Belt witnessed the largest growth?

____

14. What were the contributing factors associated with this migration?

____

____

15. What were the impacts of this migration?

____

____

16. What was the major period when the mountain states witnessed growth?

____

17. What were the contributing factors associated with this migration?

____

____

18. What were the impacts of this migration?

____

____

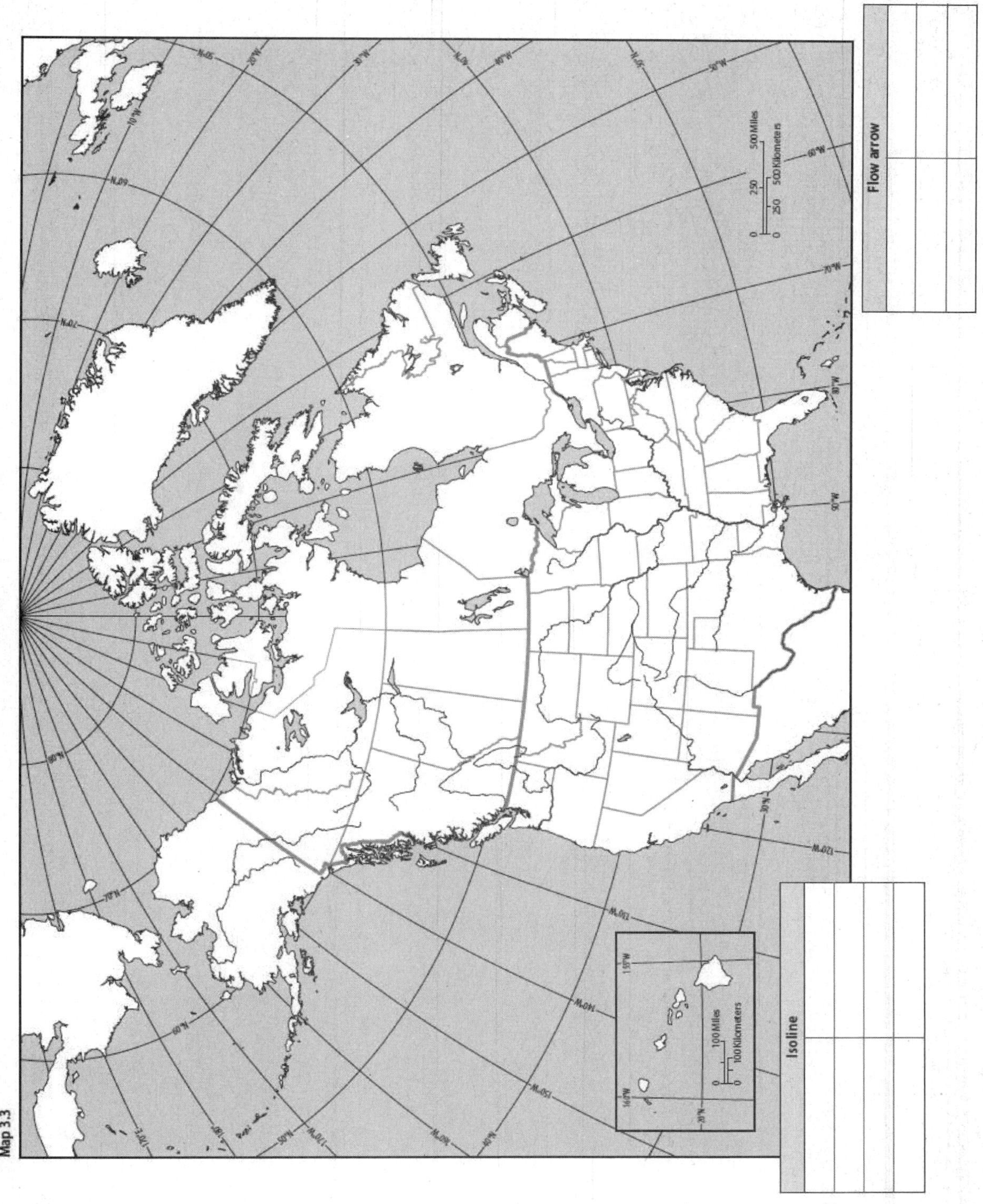
Map 3.3
0 250 500 Miles
0 250 500 Kilometers
0 100 Miles
0 100 Kilometers
Flow arrow
Isoline

Exercise Two: United States Patterns of Hispanic and Asian Residence

Using Figure 3.21 Distribution of U.S. Hispanic and Asian Household Populations, by State, 2004 (p. 66) and Mapping Workbook Maps 3.4 and 3.5, complete the following exercise.

Using Figure 3.21 "Distribution of U.S. Hispanic and Asian Household Populations, by State, 2004" (p. 66) as a reference, use a colored pencil to shade in the states that possess greater than 30 percent of the Hispanic households in the United States. Using a different colored pencil, shade in the states that possess between 18 percent and 20 percent of the Hispanic households in the United States. Using a different colored pencil, shade in the states that possess between 7 percent and 9 percent of the Hispanic households in the United States. Using a different colored pencil, shade in the states that possess between 2 percent and 5 percent of the Hispanic households in the United States. Mark your shading scheme in the map's legend.

Examine the "Asian Household by Population by State" pie chart and use a colored pencil to pattern in the states that possess greater than 30 percent of the Asian households in the United States. Using a different colored pencil, pattern in the states that possess between 18 percent and 25 percent of the Asian households in the United States. Using a different colored pencil, pattern in the states that possess between 7 percent and 10 percent of the Asian households in the United States. Using a different colored pencil, pattern in the states that possess between 2 percent and 5 percent of the Asian households in the United States. Mark your patterning scheme in the map's legend. Once you have done this, answer the following questions.

1. Which U.S. states possess the highest percentages of Hispanics?

____

____

2. What do you believe accounts for the higher percentages in these states?

____

____

3. Which U.S. states possess the lowest percentages of Hispanics?

____

____

4. What do you believe accounts for the lower percentages of Hispanics in these states?

____

____

5. Which U.S. states possess the highest percentages of Asians?

____

____

6. What do you believe accounts for the higher percentages in these states?

____

____

7. Which U.S. states possess the lowest percentages of Asians?

____

____

8. What do you think accounts for the lower percentages of Asians in these states?

9. Which states possess a correlation between larger percentages of Hispanics and larger percentages of Asians?

10. What do you think are the factors responsible for this correlation?

11. Which states possess higher percentages of Hispanics and lower percentages of Asians?

12. Do you think that there are any contributing factors associated with this reverse correlation? If so, what might those factors be?

13. Which states possess lower percentages of Hispanics and higher percentages of Asians?

14. Do you think that there are any contributing factors associated with this reverse correlation? If so, what might those factors be?

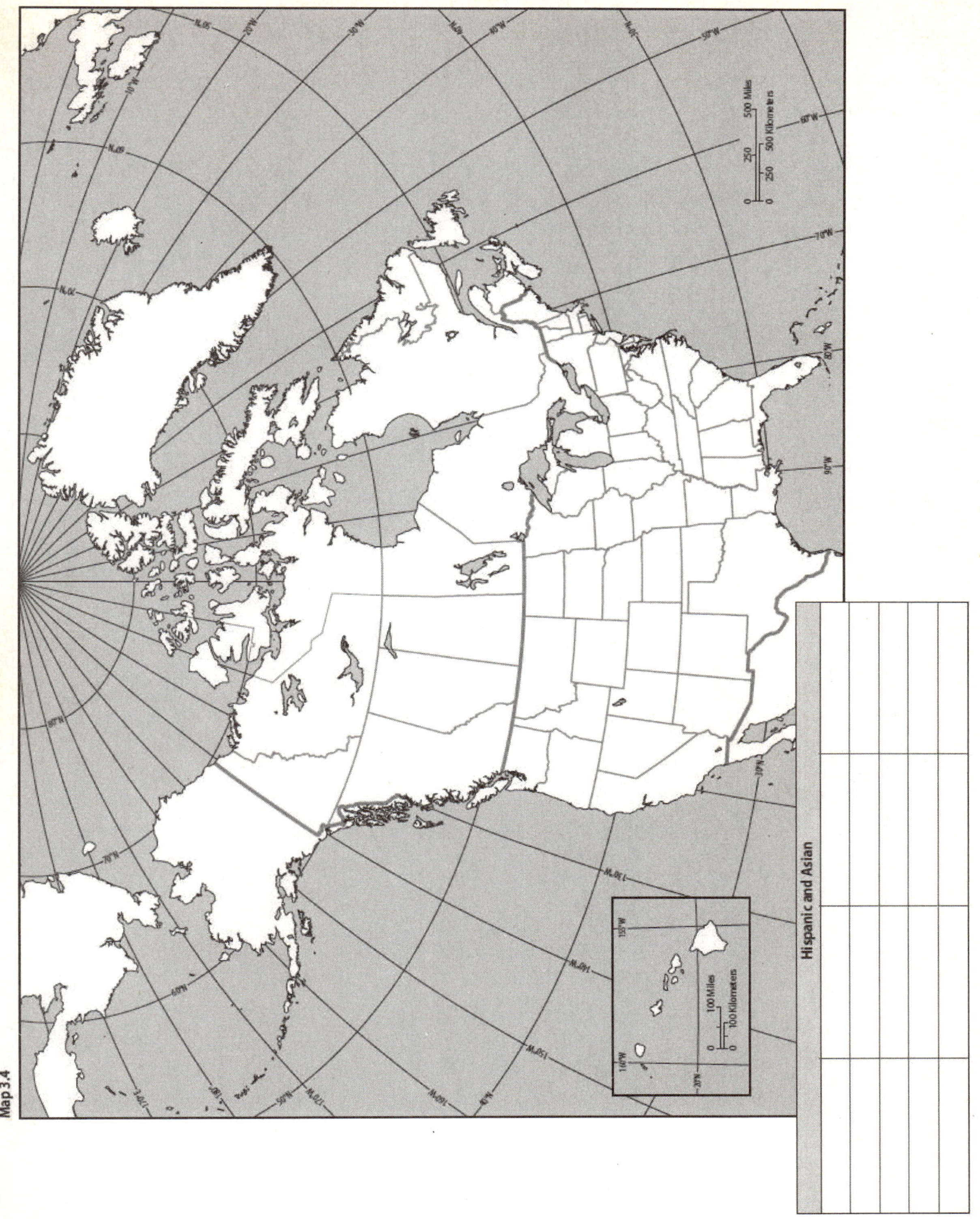
Map 3.4
Hispanic and Asian
0 250 500 Miles
0 250 500 Kilometers
0 100 Miles
0 100 Kilometers

Map 3.5

Hispanic and Asian			

Chapter Four: Latin America Mapping Workbook Exercises

Identify the following features on workbook Maps 4.1 and 4.2

Identify and label the following countries on Map 4.1

Argentina
Belize
Bolivia
Brazil
Chile
Columbia
Costa Rica
Ecuador
El Salvador
French Guiana
Guatemala
Guyana
Honduras
Mexico
Nicaragua
Panama
Paraguay
Peru
Suriname
Trinidad and Tobago
Uruguay
Venezuela

Identify and label the following cities on Map 4.1

Asunción
Bogotá
Brasilia
Buenos Aries
Caracas
Ciudad Juarez
Guatemala City
Iquitos
La Paz
La Plata
Lima
Managua
Manaus
Medellin
Mexico City
Monterrey
Montevideo
Panama City
Paramaribo
Punta Arenas
Quito
Rio de Janeiro
San Jose
San Salvador
Sal Paulo
Salvador
Santiago
Sucre
Tegucigalpa
Tijuana

Identify and label the following physical features on Map 4.2

Amazon Basin
Amazon River
Andes Mountains
Atlantic Ocean
Brazilian Highlands (Shield)
Caribbean Sea
Cordillera Occidental
Cordillera Oriental
Falkland Islands
Galapagos Islands
Guiana Highlands (Shield)
Llanos
Mato Grosso Plateau
Negro River
Orinoco River
Pacific Ocean
Parana River
Patagonian Shield
Rio de le Plata
San Francisco River
Sierra Madre Occidental
Sierra Madre Oriental
Tierra del Fuego
Uruguay River
Yucatan Peninsula

Map 4.1

0 400 800 Miles
0 400 800 Kilometers
70°W
60°W
50°W
40°W
40°N
30°N
20°N
10°N
0°
10°S
20°S
30°S
40°S
50°S
60°S
120°W
110°W
100°W
90°W
80°W
30°W

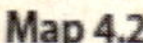

Map 4.2

Exercise One: Topography and Latin American Population Settlement

Using Figure 4.1 Latin America (physical geography) (p. 83), Figure 4.12 Climate Map of Latin America (p. 91), Figure 4.14 Population Map of Latin America (p. 93), and Mapping Workbook Map 4.3, complete the following exercise.

Using Figure 4.1 "Latin America" (physical geography) (p. 83) as a reference, use a colored pencil to shade in the low elevation regions of Latin America (0–500 feet above sea level). Using a different colored pencil, shade in the higher elevation regions of Latin America (500–2000 feet above sea level). Using a pencil that is a different color from the first two you used, shade in the highest elevation regions of Latin America (4000 + feet above sea level). Mark the shading scheme in the map's legend.

Examine Figure 4.14 "Population Map of Latin America" (p. 93) and note the major settlement patterns. Using a red or black (or a color that will contrast with the shading scheme you have already created) colored pencil, shade in the Latin American regions possessing the highest population concentration. Mark the shading scheme in the map's legend. Once you have done this, answer the following questions.

1. Describe the general topographic pattern within Mexico and Central America.

____

____

____

____

2. Describe the general topographic pattern within South America.

____

____

____

____

3. Describe the general pattern of population concentration within Mexico and Central America.

____

____

____

____

4. Describe the general pattern of population concentration within South America.

____

____

____

____

5. Overall, how does the settlement and topography correlate within Mexico and Central America?

____

____

____

6. Overall, how does the settlement and topography correlate within South America?

Examine the map you created and compare it with Figure 4.12 "Climate Map of Latin America" (p. 91).

7. In which climate(s) do we find the highest population density in Mexico and Central America?

8. Why do you think these regions are more densely inhabited?

9. In which climate(s) do we find the lowest population density within Mexico and Central America?

10. Why do you think these regions are more sparsely inhabited?

11. In which climate(s) do we find the highest population density in South America?

12. Why do you think these regions are more densely inhabited?

13. In which climate(s) do we find the lowest population density within South America?

14. Why do you think these regions are more sparsely inhabited?

Map 4.3

Population
Elevation
0 400 800 Miles
0 400 800 Kilometers
60°W
50°W
40°W
70°W
40°N
30°N
20°N
10°N
0°
10°S
20°S
30°S
40°S
50°S
60°S
80°W
90°W
100°W
110°W
120°W
30°W

Exercise Two: Latin American Economic Development

Using Table 4.2 Development Indicators (p. 107), Figure 4.33 Global Linkages: Foreign Investment and Remittances (p. 110), and Mapping Workbook Map 4.4, complete the following exercise.

Using the GNI Per Capita, PPP 2007 column in Table 4.2 "Development Indicators" (p. 107) as a reference, use a colored pencil to shade in the Latin American countries possessing an annual per capita GNI of 2,280–4,790. Using a different colored pencil, shade in the Latin American countries possessing an annual per capita GNI of 4,791–7,070. Using a different colored pencil, shade in the Latin American countries possessing an annual per capita GNI of 7,071–9,350. Using a different colored pencil, shade in the Latin American countries possessing an annual per capita GNI of 9,351–11,630. Using a different colored pencil, shade in the Latin American countries possessing an annual per capita GNI of 11,631–13,910. Mark your shading scheme in the map's legend.

Using Figure 4.33 "Global Linkages: Foreign Investment and Remittances" (p. 110) as a guide, take a ruler and draw respective proportional symbols representing the remittances per capita in U.S. dollars on their respective countries. Have a 1/2 inch (on one side) square represent greater than $300 remittances per capita, a 1/4 inch (on one side) square represent $201–$300 remittances per capita, a 1/8 inch (on one side) square represent $100–$200 remittances per capita, and a 1/16 inch (on one side) square represent less than $100 remittances per capita. Use a red or black (or a color that will contrast with the shading scheme you created above) colored pencil to shade in the proportional symbols on the map's legend.

Study the map you created and answer the below questions.

1. Which Latin American nations possess the highest annual per capita GNI?

____

____

2. Which Latin American nations possess the lowest annual per capita GNI?

____

____

3. Is there an overall regional pattern of annual per capita GNI for Latin America? If so, what is it?

____

____

____

Examine the remittances per capita proportional symbols you placed on your map.

4. Which nations receive the greatest remittances per capita?

____

____

5. Which nations receive the lowest remittances per capita?

____

____

6. Is there an overall regional pattern of remittances per capita for Latin America? If so, what is it (i.e., does the region display a sub-region that appears to receive a greater amount of remittances)?

____

____

____

Compare and contrast the remittances per capita GNI data you mapped.

7. Is there a correlation (positive or negative) between remittances and per capita GNI? If so, what is it?

____

____

____

8. If there was a correlation, what might explain it?

____

____

____

9. If there was no correlation at all, what might explain it?

____

____

____

Table

GNI	Color
2,280–4,790	
4,791–7,070	
7,071–9,350	
9,351–11,630	
11,631–13,910	

Map 4.4

0 400 800 Miles
0 400 800 Kilometers
GNI
Remittances per capita, 2006, (US $)
More than 300
201–300
100–200
Less than 100
40°N
30°N
20°N
10°N
0°
10°S
20°S
30°S
40°S
50°S
60°S
120°W
110°W
100°W
90°W
80°W
70°W
60°W
50°W
40°W
30°W

Chapter Five: Caribbean Mapping Workbook Exercises

Identify the following features on workbook Maps 5.1 and 5.2

Identify and label the following countries on Map 5.1

Anguilla
Antigua and Barbuda
Bahamas
Barbados
Belize
Bonaire
Cuba
Curacao
Dominica
Dominican Republic
French Guiana
Grenada
Guyana
Haiti
Jamaica
Martinique
Puerto Rico
St. Kitts and Nevis
St. Lucia
St. Vincent and the Grenadines
Suriname
Trinidad and Tobago

Identify and label the following cities on Map 5.1

Basseterre
Belmopan
Bridgetown
Castries
Cayenne
Georgetown
Havana
Kingston
Kingstown
Nassau
Paramaribo
Port of Spain
Port-au-Prince
Roseau
San Juan
Santo Domingo
St. George's
St. John's

Identify and label the following physical features on Map 5.2

Antillean Volcanic Arc
Atlantic Ocean
Caribbean Sea
Greater Antilles
Guiana Shield
Gulf of Mexico
Isla de Margarita
Leeward Islands
Lesser Antilles
Windward Islands

See inset
Tropic of Cancer
0 200 400 Miles
0 200 400 Kilometers
0 100 200 Miles
0 100 200 Kilometers

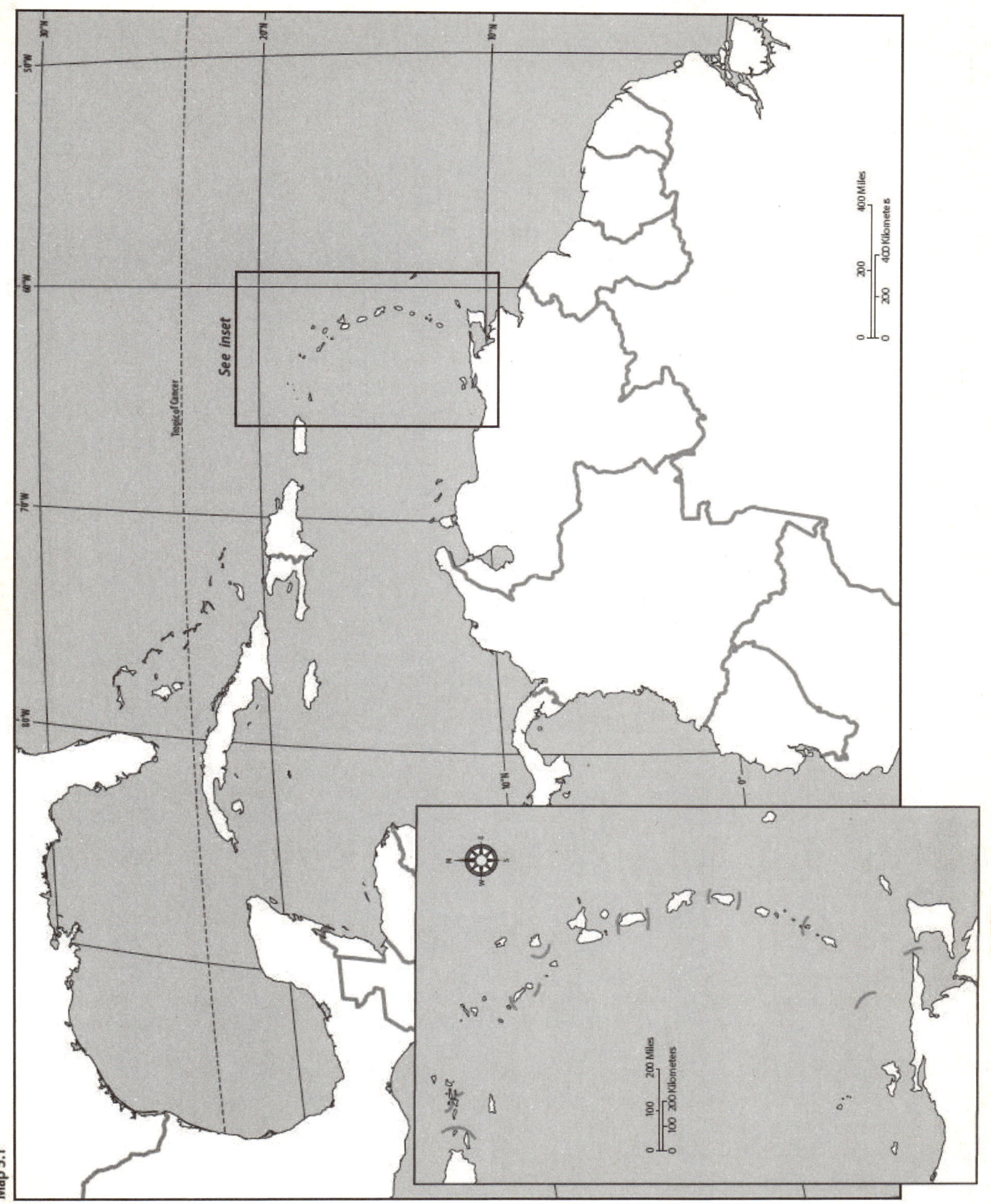

Map 5.1

30°N
20°N
10°N
0°
50°W
60°W
70°W
80°W
Tropic of Cancer
See inset
0 200 400 Miles
0 200 400 Kilometers
0 100 200 Miles
0 100 200 Kilometers

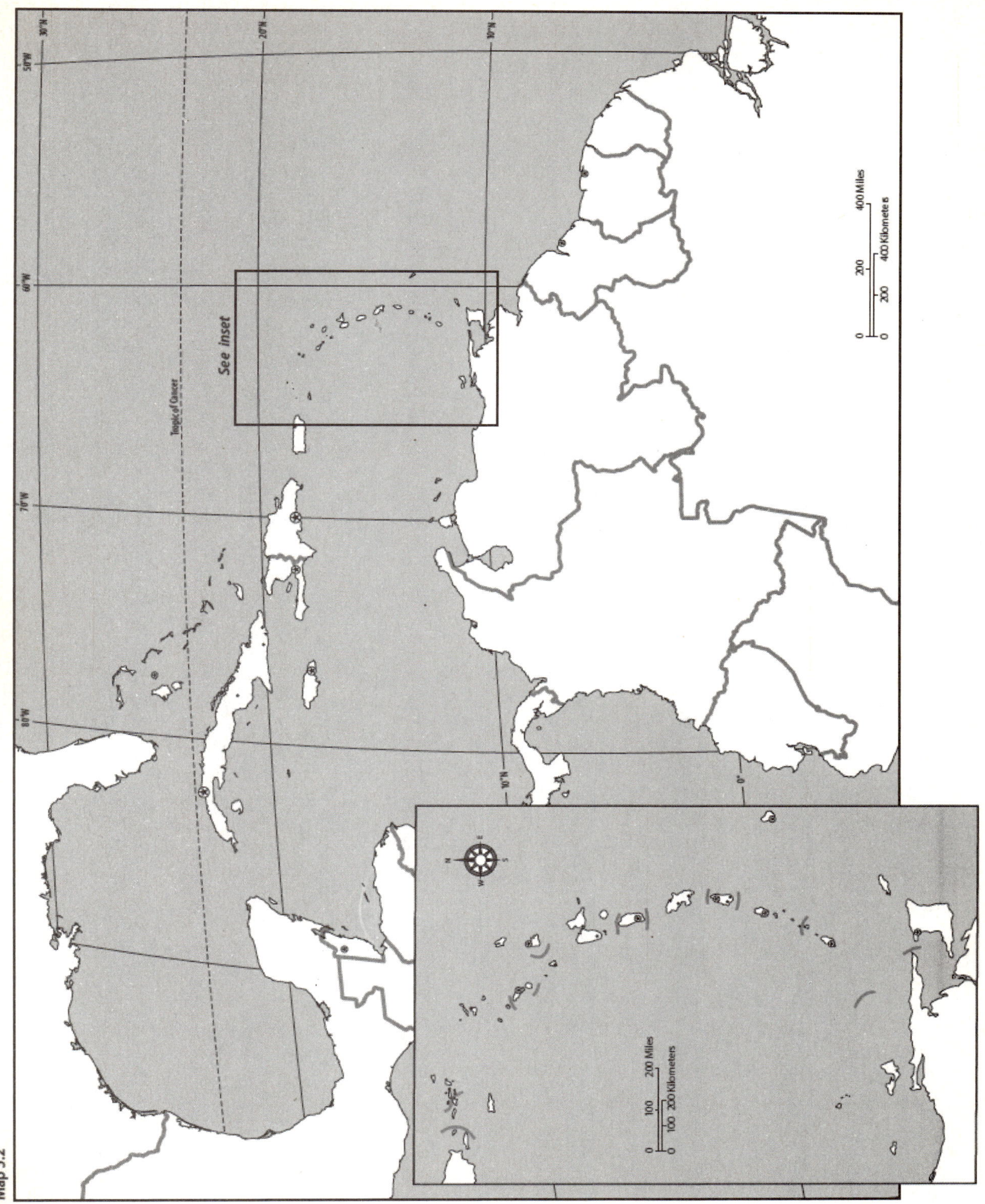

Map 5.2

Exercise One: Regional Patterns of the Caribbean Diaspora

Using Figure 5.12 Caribbean Diaspora (p. 126), Figure 5.21 Caribbean Language Map (p. 131), and Mapping Workbook Map 5.3, complete the following exercise.

Using Figure 5.21 "Caribbean Language Map" (p. 131) as a guide, use a colored pencil to shade in the Caribbean nations that possess Spanish as an official language. Using different colored pencils for each, shade in the nations possessing French, English, and Dutch as official languages. Mark your shading scheme in the map's legend. Once you do this, answer the following questions.

1. Where are the majority of the Spanish-speaking nations located within the region?

____

____

2. Where are the majority of the French-speaking nations located within the region?

____

____

3. Where are the majority of the English-speaking nations located within the region?

____

____

4. Where are the majority of the Dutch-speaking nations located within the region?

____

____

5. Based on your observations, can you identify a sub-regional pattern regarding these languages? If so, specify.

____

____

____

6. What are the factors associated with the dominance of certain European languages within the Caribbean?

____

____

7. Linguistically, how closely are these languages associated with their respective European counterparts?

____

____

8. What other linguistic influences may have come into play in the evolution or stagnation of these languages?

____

____

____

Examine Figure 5.12 "Caribbean Diaspora" (p. 126). Using a red or black (or a color that contrasts with the shading scheme you created above) colored pencil, draw flow arrows indicating the approximate origins and destinations of international and interregional migration within the Caribbean. Mark the arrow symbols in the map's legend. Once you have done this, do the below tasks and answer the questions that follow.
List the general destinations for migrants from the below Lesser Antilles, Greater Antilles and Rimland nations (if it is outside the region depicted on Figure 5.12 "Caribbean Diaspora" (p. 126) just note the migration as "international").

9. Belize ______________________________

10. French Guiana ______________________________

11. Guyana ______________________________

12. Suriname ______________________________

13. Bahamas ______________________________

14. Cuba ______________________________

15. Dominican Republic ______________________________

16. Haiti ______________________________

17. Jamaica ______________________________

18. Puerto Rico ______________________________

19. Trinidad and Tobago ______________________________

20. Smaller islands of the Lesser Antilles ______________________________

21. Does there appear to be a geographic pattern associated with migration from the Rimland nations? If so, what is it?

____

____

____

22. What do you think may account for this pattern?

____

____

____

23. Does there appear to be a geographic pattern associated with migration from the Greater Antilles nations? If so, what is it?

____

____

____

24. What do you think may account for this pattern?

25. Does there appear to be a geographic pattern associated with migration from the Lesser Antilles nations? If so, what is it?

26. What do you think may account for this pattern?

Examine the two mapped variables (language and migration) together.

27. Does there appear to be a correlation between migration origins, destinations, and language? If so, what is it?

28. If there was a correlation, how do you explain it?

Map 5.3

Languages
Flow arrow
See inset
Tropic of Cancer
80°W
70°W
60°W
50°W
30°N
20°N
10°N
0°
0 200 400 Miles
0 200 400 Kilometers
0 100 200 Miles
0 100 200 Kilometers

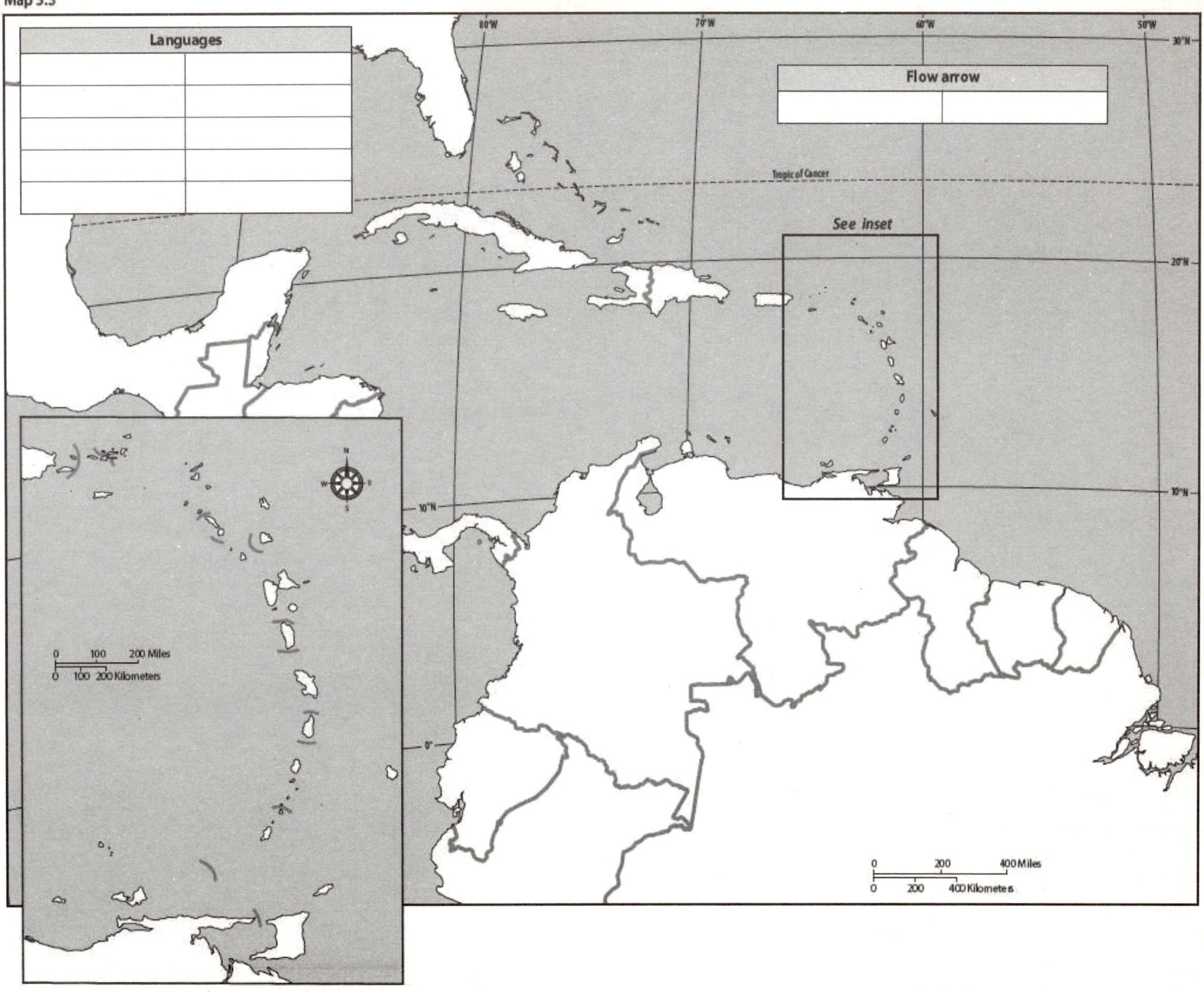

Exercise Two: Caribbean Economic Development

Using Table 5.2 Development Indicators (p. 141), Figure 5.29 Global Linkages: International Tourism in the Caribbean (p. 139), and Mapping Workbook Map 5.4, complete the following exercise.

Using the GNI Per Capita, PPP 2007 column in Table 5.2 "Development Indicators" (p. 141) as a reference, use a colored pencil to shade in the Caribbean countries possessing an annual per capita GNI of 1,050–6,560. Using a different colored pencil, shade in the Caribbean countries possessing an annual per capita GNI of 6,561–12,070. Using a different colored pencil, shade in the Caribbean countries possessing an annual per capita GNI of 12,071–17,580. Using a different colored pencil, shade in the Caribbean countries possessing an annual per capita GNI of 17,581–23,090. Using a different colored pencil, shade in the Caribbean countries possessing an annual per capita GNI of 23,091 and above. Mark your shading scheme in the map's legend. Once you have done this, answer the following questions.

1. Which Caribbean and Rimland nations possess the highest annual per capita GNI?

____

____

2. Which Caribbean and Rimland nations possess the lowest annual per capita GNI?

____

____

3. Is there an overall regional pattern of annual per capita GNI for the Caribbean and the Rimland? If so, what is it (i.e., does the region display a more and a less impoverished sub-region)?

____

____

____

4. Which sub-region appears to be the wealthiest, if any: the Greater Antilles, the Lesser Antilles, or the Rimland?

____

____

5. Can you explain the patterns of economic development, or lack thereof, in the Caribbean?

____

____

____

Using Figure 5.29 "Global Linkages: International Tourism in the Caribbean" (p. 139) as a guide, take a ruler and draw respective proportional symbols representing the tourist receipts in millions of U.S. dollars on their respective countries. Have a 1/2 inch (on one side) square represent $3000+ to $2,250 millions of tourist dollar receipts, a 1/4 inch (on one side) square represent $2,249 to $1,500 millions of tourist dollar receipts, a 1/8 inch (on one side) square represent $1,499 to $750 millions of tourist dollar receipts, and a 1/16 inch (on one side) square represent less than $750 millions of tourist dollar receipts. Use a red or black colored pencil (or a color that will contrast with the shading scheme you created above) to shade in the proportional symbols below.

Examine the tourist dollar receipts proportional symbols you placed on your map.

6. Which nations receive the greatest tourist dollar receipts?

____

____

7. Which nations receive the lowest tourist dollar receipts?

____

____

8. Is there an overall regional pattern of tourist dollar receipts for the Caribbean? If so, what is it (e.g., does the region display a sub-region that appears to receive a greater amount of tourist dollars)?

____

____

____

Compare and contrast the tourist dollars with the annual per capita GNI data you mapped.

9. Is there a correlation (positive or negative) between tourist dollars and per capita GNI? If so, what is it? If so, what might explain it?

____

____

____

____

____

____

GNI	Color
1,050–6,560	
6,561–12,070	
12,071–17,580	
17,581–23,090	
23,091 and above	

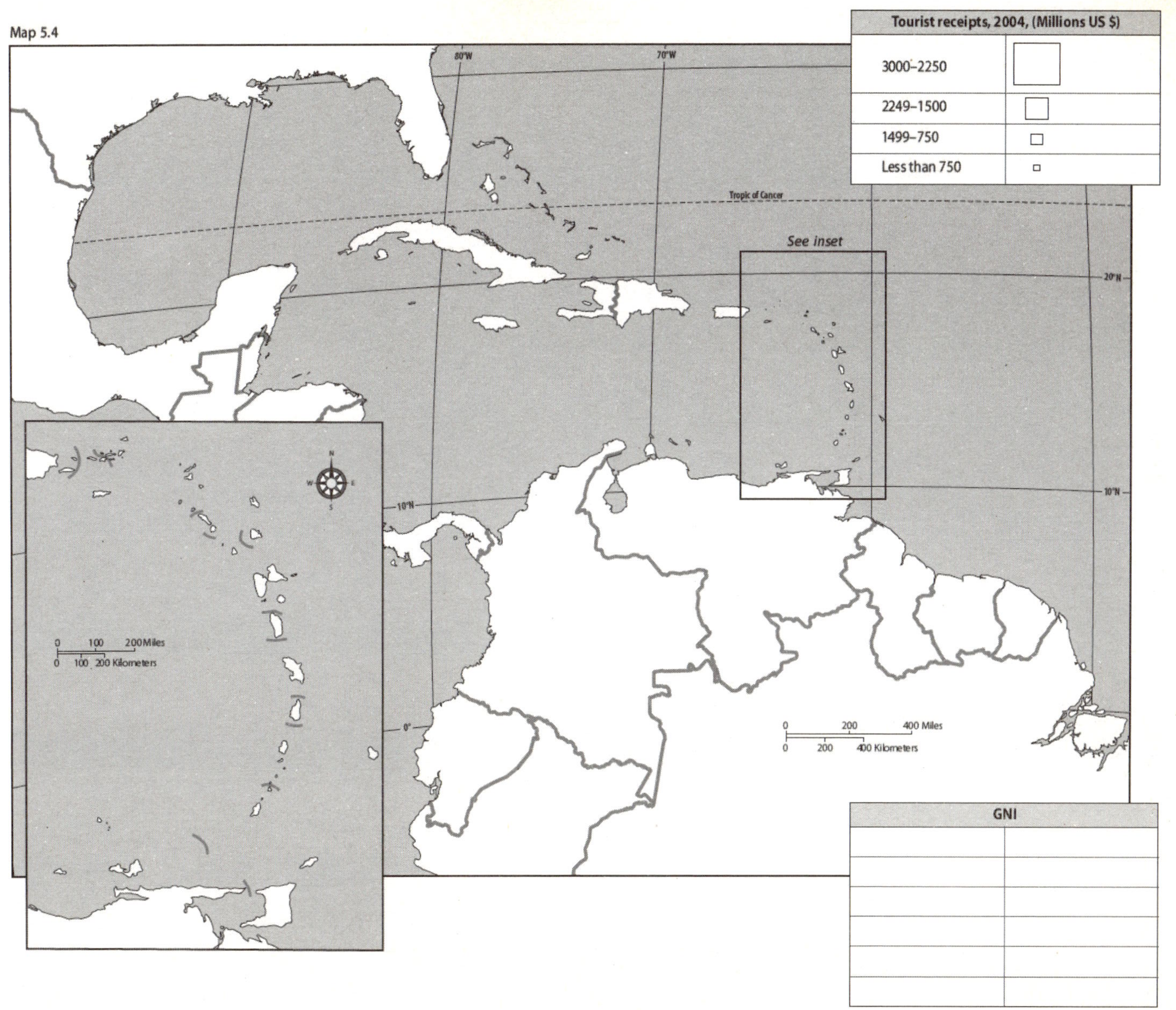

Map 5.4
Tourist receipts, 2004, (Millions US $)
3000–2250
2249–1500
1499–750
Less than 750
80°W
70°W
Tropic of Cancer
See inset
20°N
10°N
0°
N
W
E
S
0 100 200 Miles
0 100 200 Kilometers
0 200 400 Miles
0 200 400 Kilometers
GNI

Chapter Six: Sub-Saharan African Mapping Workbook Exercises

Identify the following features on workbook Maps 6.1 and 6.2

Identify and label the following countries on Map 6.1

Angola
Benin
Botswana
Burkina Faso
Burundi
Cabinda (Angola)
Cameroon
Cape Verde
Central African Republic
Chad
Comoros Islands
Democratic Republic of the Congo
Djibouti
Equatorial Guinea
Eritrea
Ethiopia
Gabon
Gambia
Ghana
Guinea
Guinea-Bissau
Ivory Coast
Kenya
Lesotho
Liberia
Madagascar
Malawi
Mali
Mauritania
Mauritius
Mozambique
Namibia
Niger
Nigeria
Republic of the Congo
Rwanda
Sao Tome and Principe
Senegal
Seychelles
Sierra Leone
Somalia
South Africa
Sudan
Swaziland
Tanzania
Togo
Uganda
Western Sahara
Zambia
Zimbabwe

Identify and label the following cities on Map 6.1

Abuja
Accra
Addis Ababa
Antananarivo
Asmara
Bamako
Bangui
Banjul
Bissau
Brazzaville
Bujumbura
Cape Town
Conakry
Dakar
Dar es Salaam
Djibouti
Freetown
Gaborone
Harare
Johannesburg
Kampala
Khartoum
Kigali
Kinshasa
Lagos
Libreville
Lilongwe
Lome
Luanda
Lusaka
Malabo
Maputo
Maseru
Mogadishu
Monrovia
Moroni
N'Djamena
Nairobi
Niamey
Nouakchott
Ouagadougou
Port Louis
Porto-Novo
Praia
Pretoria
Sao Tome
Tombocutou
Victoria
Walvis Bay
Windhoek
Yamoussoukro
Yaounde

Identify and label the following physical features on Map 6.2

Adamawa Highlands
Atlantic Ocean
Benue River
Blue Nile River
Caprivi Strip
Congo Basin
Congo River
Ethiopian Highlands
Great Escarpment
Great Rift Valley
Gulf of Aden
Horn of Africa
Indian Ocean
Kalahari Desert
Kasai River
Lake Chad
Lake Nyasa
Lake Tana
Lake Tanganyika
Lake Turkana
Lake Victoria
Limpopo River
Mozambique Channel
Mt. Kenya
Mt. Kilimanjaro
Namib Desert
Niger River
Nile River
Orange River
Red Sea
Reunion
Sahara Desert
Sahel
Senegal River
Ubangi River
White Nile River
Zambezi River

Equator
Tropic of Cancer
Tropic of Capricorn
0 400 800 Miles
0 400 800 Kilometers

Map 6.1

0 400 800 Miles
0 400 800 Kilometers
Equator
Tropic of Capricorn
Tropic of Cancer

Map 6.2

Exercise One: Regional Patterns of the Language and Colonization

Using Figure 6.25 African Language Groups and Official Languages (p. 164), Figure 6.33 European Colonization in 1913 (p.171), and Mapping Workbook Map 6.3, complete the following exercise.

Using Figure 6.25 "African Language Groups and Official Languages" (p. 164) as a reference, use a colored pencil to shade in the African nations that have Amharic as their official language. Using a different colored pencil to represent each language, shade in the official languages of each of the African nations. Mark your shading scheme in the map's legend. Once you have done this, answer the below questions.

1. In which Sub-Saharan African nation(s) do we find Amharic as the official language?

____

____

2. In which Sub-Saharan African nation(s) do we find Arabic as the official language?

____

____

3. In which Sub-Saharan African nation(s) do we find English as the official language?

____

____

4. In which Sub-Saharan African nation(s) do we find French as the official language?

____

____

5. In which Sub-Saharan African nation(s) do we find Portuguese as the official language?

____

____

6. In which Sub-Saharan African nation(s) do we find Somali as the official language?

____

____

7. In which Sub-Saharan African nation(s) do we find Spanish as the official language?

____

____

8. In which Sub-Saharan African nation(s) do we find Swahili as the official language?

____

____

9. In which Sub-Saharan African nation(s) do we find no official language or a dominance of multiple languages?

____

____

10. Does there appear to be a regional pattern associated with the languages within Sub-Saharan Africa? If so, what is it?

____

____

____

11. If there is a regional pattern, what do you think the cause of it is?

____

____

____

Examine Figure 6.33 "European Colonization in 1913" (p. 171). Draw a 1/4 inch square over each Sub-Saharan African nation. Use a colored pencil to pattern each square to represent each of the major European colonizing nations that are listed in the legend of Figure 6.33 "European Colonization in 1913" (p. 171) (try to select a color that will contrast with the colors you used to depict official languages). Mark your patterning scheme in the map's legend. Once you have done this, answer the below questions.

Colonizing Nation	**Symbol**	**Colonizing Nation**	**Symbol**
Belgium		Italy	
Britain		Portugal	
France		Spain	
Germany		Independent	

12. Which Sub-Saharan African nation(s) today have a Belgian colonial legacy?

____

____

13. Which Sub-Saharan African nation(s) today have a British colonial legacy?

____

____

14. Which Sub-Saharan African nation(s) today have a French colonial legacy?

____

____

15. Which Sub-Saharan African nation(s) today have a German colonial legacy?

____

____

16. Which Sub-Saharan African nation(s) today have an Italian colonial legacy?

____

____

17. Which Sub-Saharan African nation(s) today have a Portuguese colonial legacy?

18. Which Sub-Saharan African nation(s) today have a Spanish colonial legacy?

19. Which Sub-Saharan African nation(s) have remained un-colonized?

20. Is there a regional pattern associated with European colonization? If so, what is it?

21. How would you explain such a regional pattern?

Compare and contrast the official languages within the various Sub-Saharan African nations with the colonial legacy symbols you drew.

22. Is there a correlation between the European colonizers and official languages within these nations? Please provide specific examples.

23. What do you believe to be the cause of this correlation?

____

____

____

____

____

____

24. Are there any outliers to this situation? For example, are there Sub-Saharan African nations that possess a completely different official language from that of their colonizers? Please provide specific examples.

____

____

____

____

____

____

25. What do you think are the causes of these outliers?

____

____

____

____

____

____

Map 6.3

Languages
Colonization
0
400
800 Miles
0
400
800 Kilometers
Equator
Tropic of Cancer
Tropic of Capricorn
30°N
20°N
10°N
0°
10°S
20°S
30°S
30°W
20°W
10°W
0°
10°E
20°E
30°E
40°E
50°E

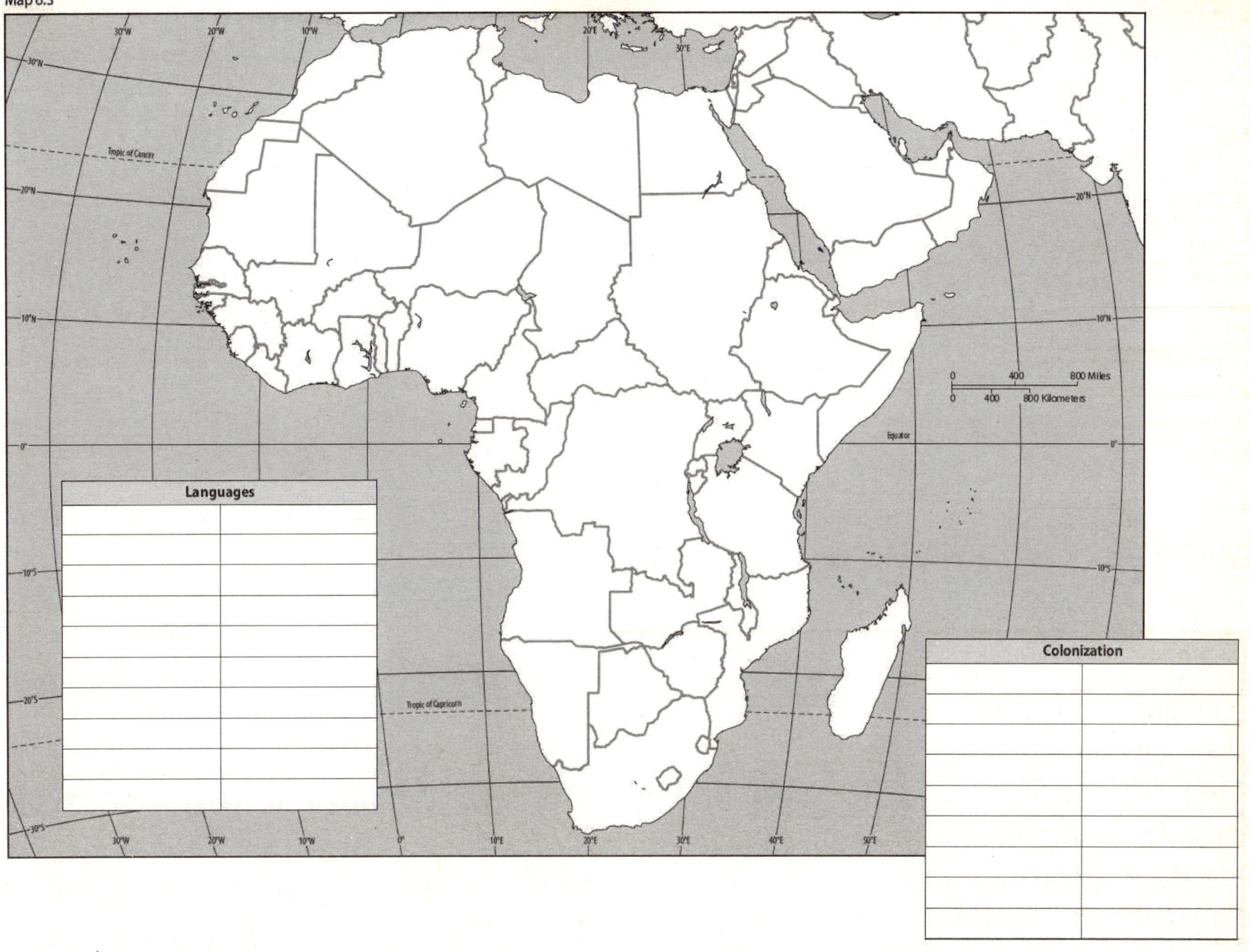

Exercise Two: Patterns of Change Under Age 5 Mortality in Sub-Saharan Africa

Using Table 6.2 Development Indicators (p. 176) and Mapping Workbook Map 6.4, complete the following exercise.

Examine Table 6.2 "Development Indicators" (p. 176) and in the table below record the under age 5 mortality rates for 1990 and 2007 in their respective columns. Once you do this, determine the difference between the 1990 and 2007 rates by subtracting the 2007 under age 5 mortality rates from the 1990 under age 5 mortality rates. Record the difference in the "Rate Difference" column in the table. Be certain to indicate whether there was a decrease (use a minus "-" symbol to indicate a decrease) or an increase (use a "+" symbol to indicate an increase) in the under age 5 mortality rate.

Once you have done this, use a colored pencil to shade in those countries that have experienced a decrease in their under age 5 mortality rates. Specifically, use one color to indicate countries that have experienced a net decrease of 128–96 in their under age 5 mortality rate, a different colored pencil to indicate countries that have experienced a net decrease of 95–64 in their under age 5 mortality rate, a different colored pencil to indicate countries that have experienced a net decrease of 63–32 in their under age 5 mortality rate, and a different colored pencil to indicate countries that have experienced a net decrease of 31–1 in their under age 5 mortality rate. Also, shade in the respective colors in the legend below.

Once you have done this, use different colored pencils to shade in those countries that have experienced an increase in their under age 5 mortality rates. Specifically, use one color to indicate countries that have experienced a net increase of 36–24 in their under age 5 mortality rate, a different colored pencil to indicate countries that have experienced a net increase of 23–12 in their under age 5 mortality rate, and a different colored pencil to indicate countries that have experienced a net increase of 12–1 in their under age 5 mortality rate. Also, shade in the respective colors in the legend below.

Rate of Decrease Under Age 5 Mortality	Color	Rate of Increase Under Age 5 Mortality	Color
-128–96		+36–24	
-95–64		+23–12	
-63–32		+12–1	
-31–1			

Once you have done the above, answer the following questions.

1. Which Sub-Saharan African nations have experienced the largest rates of decrease in their under age 5 mortality between 1990 and 2007?

____

____

____

2. Does there appear to be a geographic pattern associated with those Sub-Saharan African nations that have experienced such a decrease?

____

____

____

3. If there is a geographic pattern, what could possibly explain it?

4. Closely examine the data for the nations that have experienced the lowest rates of change in the decrease of their under age 5 mortality (not those that have experienced a gain, but rather those that have experienced rates of only -31–1) between 1990 and 2007. In regard to these nations, what can be stated about their lower rates of decrease?

5. Now closely examine the data of those nations that have experienced the highest rate of decrease (-128–96) in their under age 5 mortality between 1990 and 2007. In regard to these nations, what can be stated about their higher rates of decrease?

6. Now contrast the nations that experienced the lowest rate of decrease (-31–1) with the data of those nations that experienced the greatest rate of decrease (-128–96) in under age 5 mortality between 1990 and 2007. Is a small decrease in the under age 5 mortality in a given nation as compared to a large decrease in the under age 5 mortality rate in a given nation a comparatively better situation? Explain.

7. Which Sub-Saharan African nations have experienced an overall increase with their rates of under age 5 mortality between 1990 and 2007? Please specify.

8. Does there appear to be a geographic pattern associated with those Sub-Saharan African nations that have experienced such an increase?

9. If there is a geographic pattern, what could possibly explain it?

10. As compared to those Sub-Saharan African nations that experienced an overall decrease in their under age 5 rates of mortality between 1990 and 2007, is there a commonality among those that experienced increased rates of under age 5 mortality during the same period? More specifically, what could be the possible causes of increased rates of under age 5 mortality within these nations during this period?

Country	Under Age 5 Mortality Rate, 1990	Under Age 5 Mortality Rate, 2007	Rate Difference (=1990–2007 data)
Angola			
Benin			
Botswana			
Burkina Faso			
Burundi			
Cameroon			
Cape Verde			
Central African Rep.			
Chad			
Comoros			
Congo			
Dem. Rep. of the Congo			
Djibouti			
Equatorial Guinea			
Eritrea			
Ethiopia			
Gabon			
Gambia			
Ghana			
Guinea			
Guinea-Bissau			
Kenya			
Lesotho			
Liberia			
Madagascar			
Malawi			
Mali			
Mauritania			
Mauritius			

Mozambique			
Namibia			
Niger			
Nigeria			
Rwanda			
Sao Tome and Principe			
Senegal			
Seychelles			
Sierra Leone			
Somalia			
South Africa			
Sudan			
Swaziland			
Tanzania			
Togo			
Uganda			
Zambia			
Zimbabwe			

Under Age 5 Mortality
Equator
Tropic of Capricorn
Tropic of Cancer
0 400 800 Miles
0 400 800 Kilometers

Map 6.4

Chapter Seven: Southwest Asia and North Africa Mapping Workbook Exercises

Identify the following features on workbook Maps 7.1 and 7.2

Identify and label the following countries on Map 7.1

Algeria
Bahrain
Egypt
Iran
Iraq
Israel
Jordan
Kuwait
Lebanon
Libya
Morocco
Oman
Qatar
Saudi Arabia
Sudan
Syria
Tunisia
Turkey
U.A.E.
Western Sahara
Yemen

Identify and label the following cities on Map 7.1

Abu Dhabi
Algiers
Amman
Ankara
Baghdad
Basra
Beirut
Bursa
Cairo
Damascus
Doha
El Aaiun
Istanbul
Jerusalem
Kuwait City
Manama
Makkah
Mosul
Muscat
Rabat
Riyadh
San'a
Tehran
Tel Aviv
Tripoli
Tunis

Identify and label the following physical features on Map 7.2

Ahaggar Mountains
Anatolian Plateau
Arabian Peninsula
Atlantic Ocean
Atlas Mountains
Black Sea
Blue Nile River
Caspian Sea
Dead Sea
Elburz Mountains
Euphrates River
Gulf of Aden
Indian Ocean
Iranian Plateau
Jordan River
Lake Nasser
Libyan Desert
Maghreb
Mediterranean Sea
Nile River
Persian Gulf
Red Sea
Rub-Al-Khali
Sinai
Socotra
Strait of Gibraltar
Straits of Hormuz
Suez Canal
Tigris River
White Nile River
Yemen Highlands
Zagros Mountains

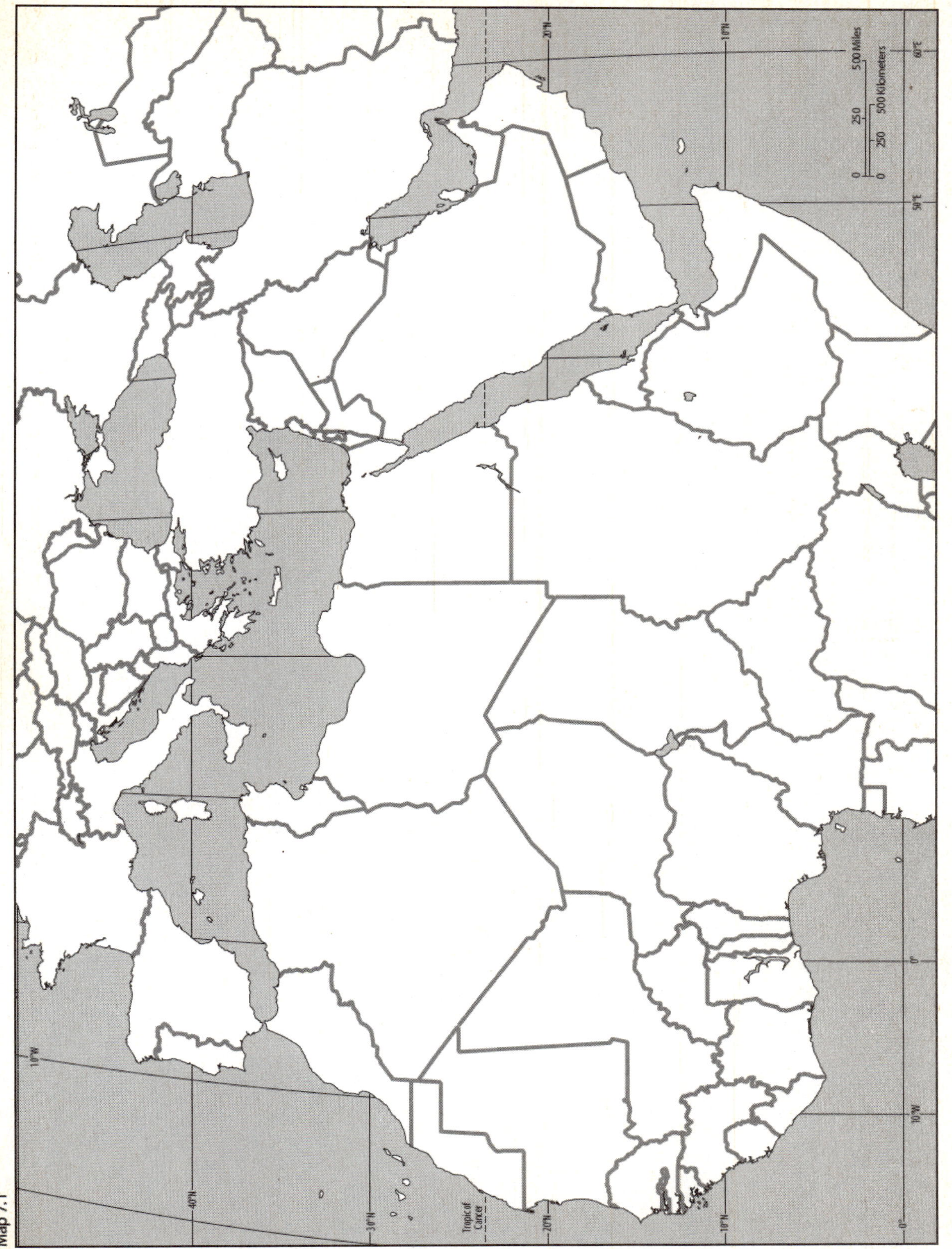
Map 7.1
0 250 500 Miles
0 250 500 Kilometers
60°E
50°E
20°N
10°N
0°
10°W
10°W
40°N
30°N
Tropic of Cancer
20°N
10°N
0°

0 250 500 Miles
0 250 500 Kilometers
60°E
50°E
20°N
10°N
0°
10°W
40°N
30°N
Tropic of Cancer

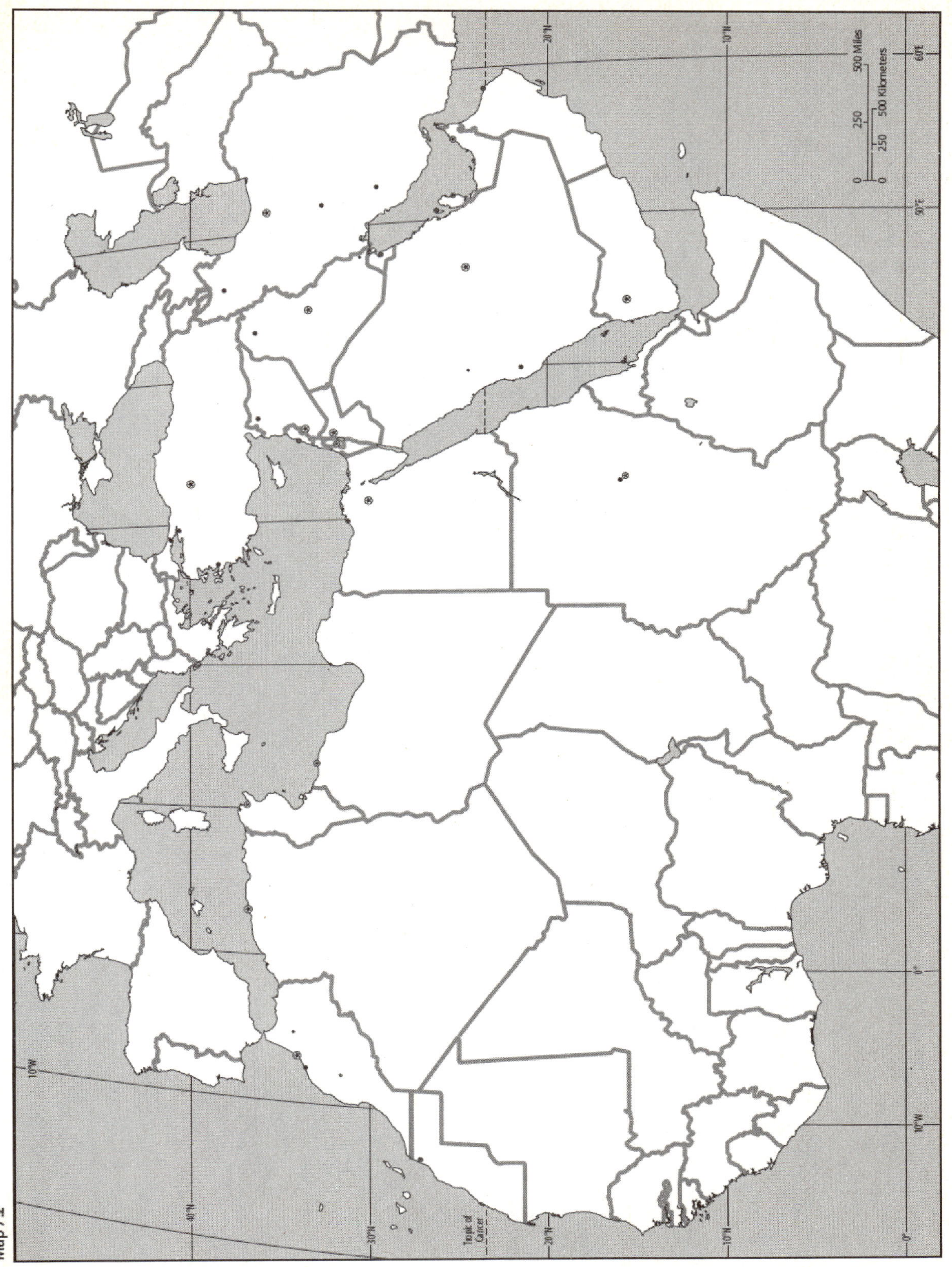

Map 7.2

Exercise One: Climate and Southwest Asian and North African Agriculture

Using 7.1 Southwest Asia and North Africa (physical features) (pp. 186-187), Figure 7.11 Climate Map of Southwest Asia and North Africa (p.193), Figure 7.17 Agricultural Regions of Southwest Asia and North Africa (p.197), and Mapping Workbook Map 7.3, complete the following exercise.

Using Figure 7.11 "Climate Map of Southwest Asia and North Africa" (p.193) as a reference, use a colored pencil to shade in the regions that possess a BWh climate. Using a different color to represent each climate type, shade in the other major climates within the region. Mark your shading scheme in the map's legend. Once you have done this, answer the following questions.

1. What are the major climates of Southwest Asia and North Africa?

____

____

2. What are the general locations of these climates?

____

____

Using 7.1 "Southwest Asia and North Africa" (physical features) (pp. 186-187) as a reference, use a colored pencil to draw symbols representing mountains, deserts, rivers, and other physical features. Use different symbols for each category of feature. Mark the symbols in the map's legend. Once you have done this, answer the below questions.

3. What are the major physical features of Southwest Asia and North Africa?

____

____

4. What are the general locations of these physical features?

____

____

Examine the physical features you drew and compare them with the major climates that you shaded in on your map.

5. Do you see a correlation between the location of some of these physical features and the climate types within Southwest Asia and North Africa? If so, what are they? Please provide specific examples.

____

____

____

____

6. How do you think that climate is affected by some of these physical features? More specifically, what are some of the controls that the physical features within this region have on climate? Explain.

____

____

____

____

Using Figure 7.17 "Agricultural Regions of Southwest Asia and North Africa" (p.197) as a reference use a red or black (or a color that contrasts with the shading scheme you created for climate) colored pencil to shade in the major agricultural types within the region. Use different patterns for pastoral nomadism, oasis and irrigated agriculture, and dry farming (with some irrigation). Mark the shading/patterning scheme in the map's legend. Once you have done this, answer the following questions.

7. What is (are) the location(s) of pastoral nomadism within Southwest Asia and North Africa?

8. What is (are) the location(s) of oasis and irrigated agriculture within Southwest Asia and North Africa?

9. What is (are) the location(s) of dry farming (with some irrigation) within Southwest Asia and North Africa?

Examine the agricultural regions and their location relative to the climate types and physical features you shaded in on the map.

10. What physical features and climate correlate with pastoral nomadism, if any?

11. How would you explain the correlation between this agricultural activity and the physical features and climate within the region where it prevails? If there was no correlation, how would you explain the lack of one?

12. What physical features and climate correlate with oasis and irrigated agriculture, if any?

13. How would you explain the correlation between this agricultural activity and the physical features and climate within the region where it prevails? If there was no correlation, how would you explain the lack of one?

14. What physical features and climate correlate with dry farming (with some irrigation), if any?

15. How would you explain the correlation between this agricultural activity and the physical features and climate within the region where it prevails? If there was no correlation, how would you explain the lack of one?

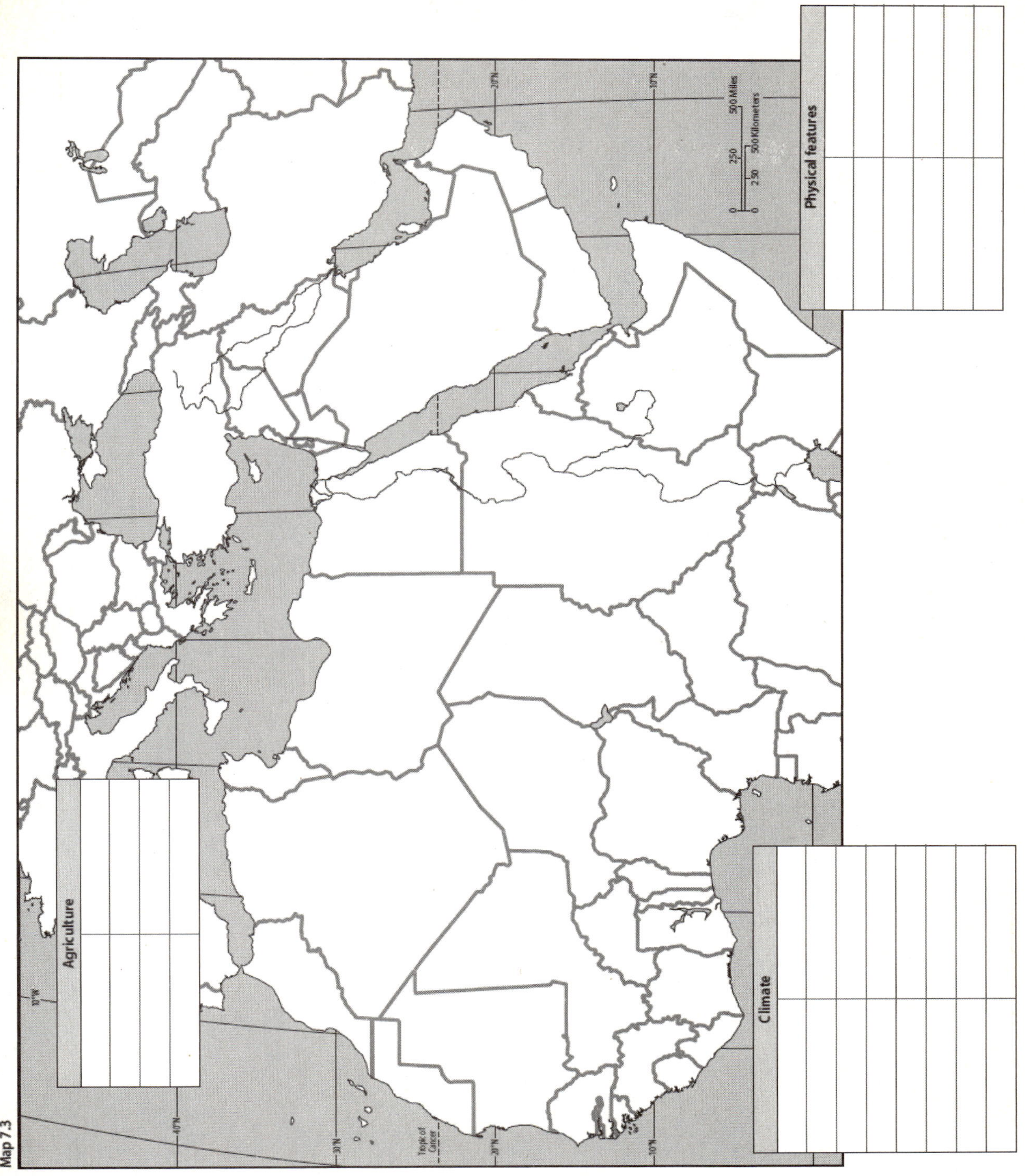
Map 7.3
Agriculture
Climate
Physical features
0 250 500 Miles
0 250 500 Kilometers
Tropic of Cancer
10°W
40°N
30°N
20°N
10°N

Exercise Two: Changing Political Borders within Israel

Using Figure 7.32 Evolution of Israel (p. 208), Figure 7.33 West Bank (p. 209), Figure 7.35 Israeli Security Barrier (p. 210), and Mapping Workbook Map 7.4, complete the following exercise.

Using Figure 7.32 "Evolution of Israel" (p. 208) as a reference, use a colored pencil to draw a line around the region that was established by the British between 1922 and 1948. Label this line "1922-1948." Using different colored pencils to represent each isochron (line of equal time), draw lines depicting the evolution of Israel under the UN partitioning plan of 1947, Israel after 1948-1949, and the region since the 1967 war. Label each line with its respective date. Mark these lines in the map's legend. Once you have done this, answer the following questions.

1. Describe the relative areas and comparative geographic locations of the Arab Muslim and Jewish states that were created under the UN partitioning plan.

2. Generally describe the circumstances that brought about these territorial changes.

3. How might have the size and the locations of these states had an effect on the access to resources and on the political relations between the citizens of these two states?

4. Describe the relative areas and comparative geographic locations of Israel and Jordan after 1948–1949.

5. Generally describe the circumstances that brought about these territorial changes.

6. How might have the size and the locations of these states had an effect on the access to resources and on the political relations between the citizens of these two states, and with other inhabitants within the bordering nations?

7. Describe the relative areas and comparative geographic locations of Israel, Jordan, Egypt and Lebanon after the 1967 war.

8. Generally describe the circumstances that brought about these territorial changes.

9. How might have the size and the locations of these states had an effect on the access to resources and on the political relations between the citizens of these two states, and with other inhabitants within the bordering nations?

Using Figure 7.33 "West Bank" (p. 209) as a reference, use a red or black colored pencil to draw in the general location of West Bank Palestinian settlements that are under full or partial Palestinian control. Mark this symbol in the map's legend. Once you have done this, answer the below questions.

10. Describe the general location of the Palestinian settlements within the West Bank. Are they clustered, dispersed, etc.?

11. How do you think the locations of these settlements have affected the political success and unity of the Palestinians within the West Bank? Explain.

Using Figure 7.35 "Israeli Security Barrier" (p. 210) as a reference, use a colored pencil (one that will contrast with the other colors you have already used) to draw a line indicating the location of the security barrier and the general location of the West Bank Israeli settlements. Mark this symbol in the map's legend. Once you have done this, answer the below questions.

12. Describe the general location of the Israeli settlements within the West Bank. Are they clustered, dispersed, etc.?

13. How do you think the locations of these settlements have affected the political success and unity of the Israelis within the West Bank? Explain.

14. Compare the location of the Palestinian West Bank settlements and the location of the Israeli West Bank settlements. Describe the relative locations of these settlements.

15. How might their locations affect the political success and unity of the Palestinians and Israelis within the West Bank?

16. Examine the security barrier line you drew on your map. How well does the line correspond to the locations of the Palestinian and Israeli settlements within the West Bank? Specifically, describe its relative location to these settlements.

17. Considering such issues as the movement of people and goods into and out of, as well as within, the West Bank, how might this security barrier affect these settlements and overall life within the region?

18. What has been, and what will be, the effect of the continued construction of the settlement barrier on the political situation within the region? Explain.

Map 7.4

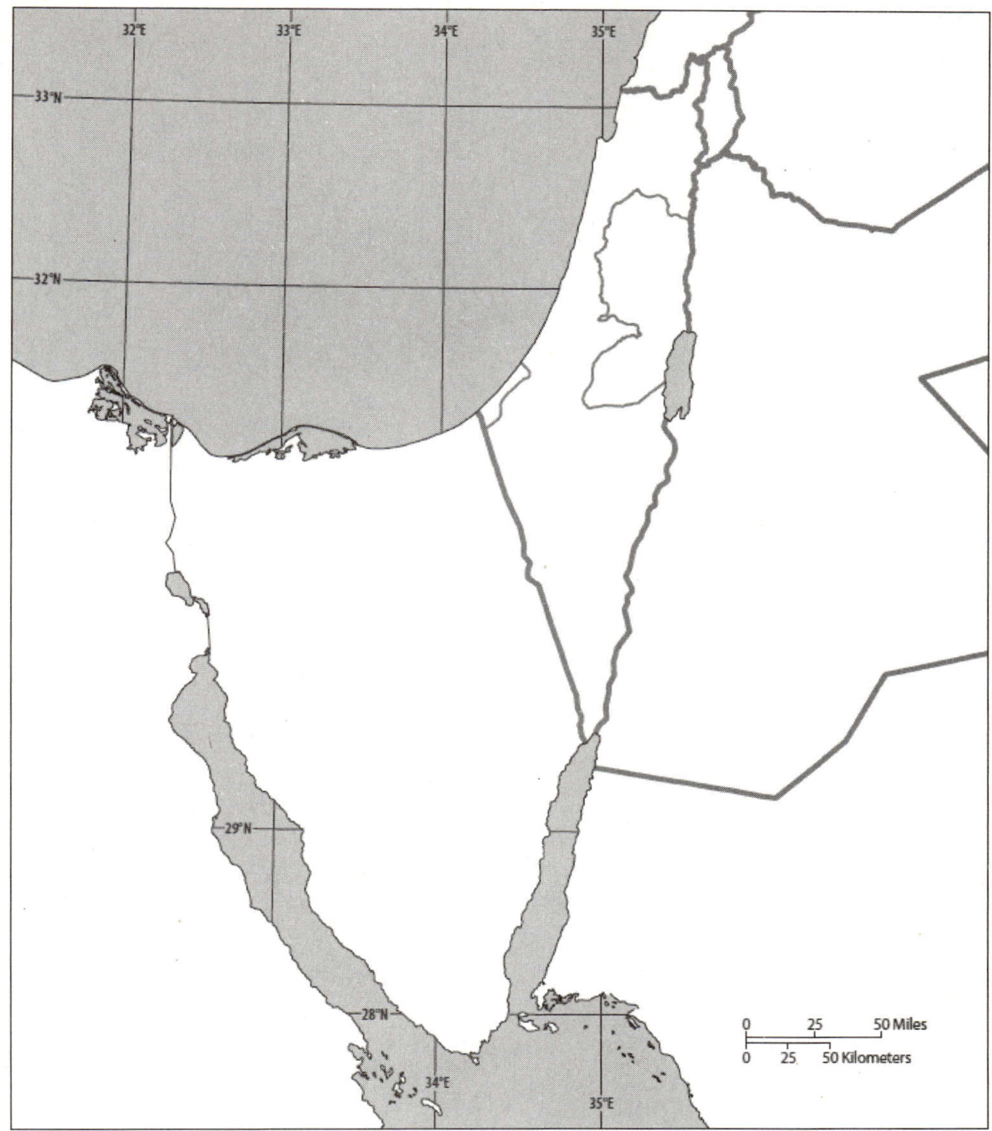

Isochron	

Security Barrier and Settlements	

Chapter Eight: Europe Mapping Workbook Exercises

Identify the following features on workbook Maps 8.1 and 8.2

Identify and label the following countries on Map 8.1

Albania
Andorra
Austria
Belgium
Bosnia & Herzegovina
Bulgaria
Croatia
Czech Republic
Denmark
Estonia
Finland
France
Germany
Greece
Hungary
Iceland
Ireland
Italy
Kosovo
Latvia
Liechtenstein
Lithuania
Luxembourg
Macedonia
Malta
Monaco
Montenegro
Netherlands
Norway
Poland
Portugal
Romania
San Marino
Serbia
Slovakia
Slovenia
Spain
Sweden
Switzerland
United Kingdom
Vatican City

Identify and label the following cities on Map 8.1

Amsterdam
Athens
Belfast
Belgrade
Berlin
Bern
Bratislava
Brussels
Bucharest
Budapest
Copenhagen
Dublin
Glasgow
Helsinki
Lisbon
Ljubljana
London
Luxembourg
Madrid
Oslo
Paris
Podgorica
Prague
Pristina
Reykjavik
Riga
Rome
Sarajevo
Skopje
Sofia
Stockholm
Tallinn
Tirana
Valletta
Vienna
Vilnius
Warsaw
Zagreb

Identify and label the following physical features on Map 8.2

Adriatic Sea
Aegean Sea
Alps
Appenines Mountains
Atlantic Ocean
Balearic Islands
Baltic Sea
Bay of Biscay
Carpathians
Corsica
Crete
Danube River
Ebro River
Elbe River
Faroe Islands
Garonne River
Loire River
Mediterranean Sea
North Sea
North European Lowland
Norwegian Sea
Oder River
Orkney Islands
Po River
Pyrenees
Rhine River
Rhone River
Sardinia
Seine River
Shetland Islands
Sicily
Strait of Gibraltar
Tagus River
Thames River
Vistula River

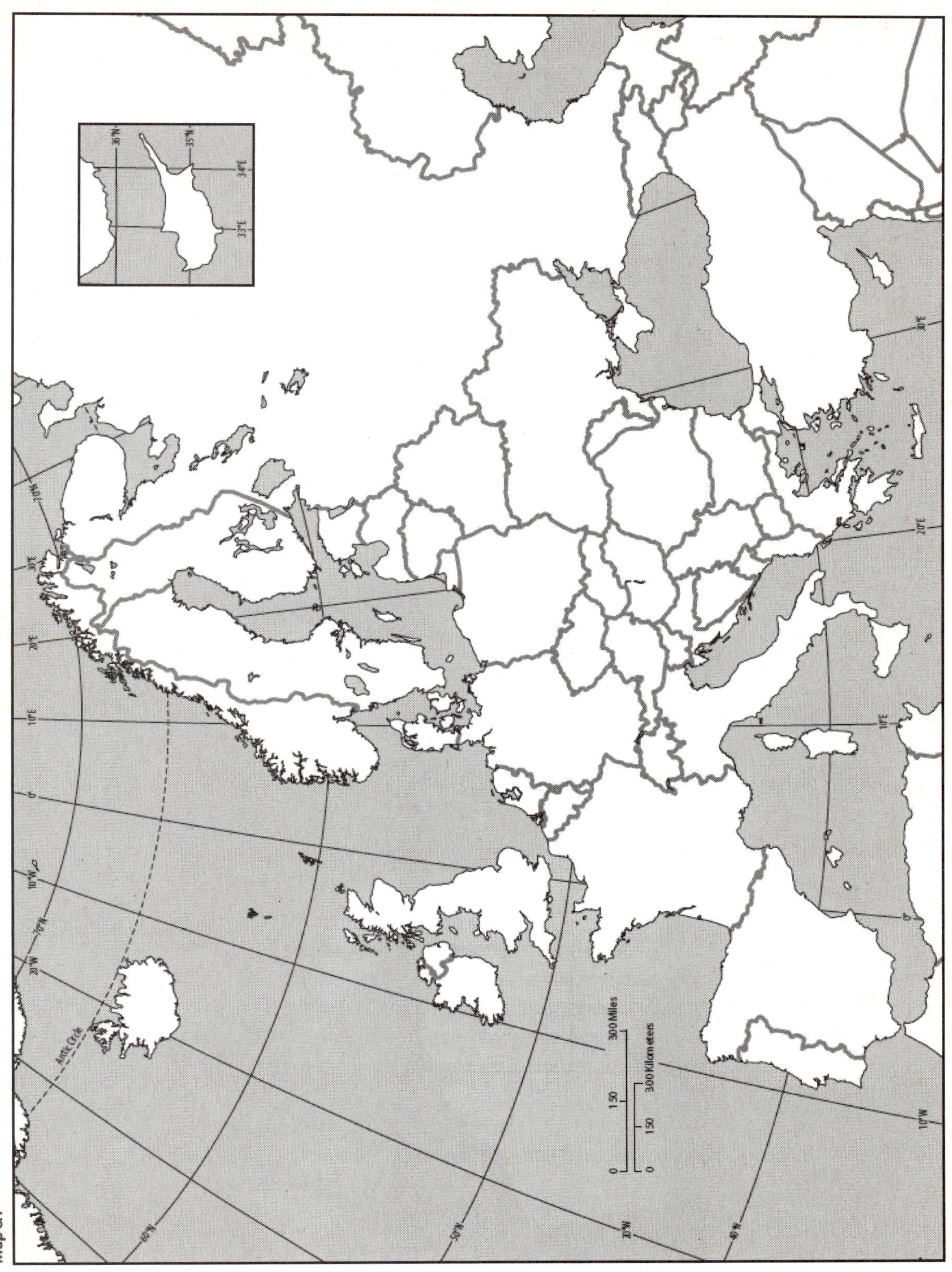
Map 8.1
300 Miles
150
0
300 Kilometers
150
0
Arctic Circle
36°N
35°N
34°E
33°E

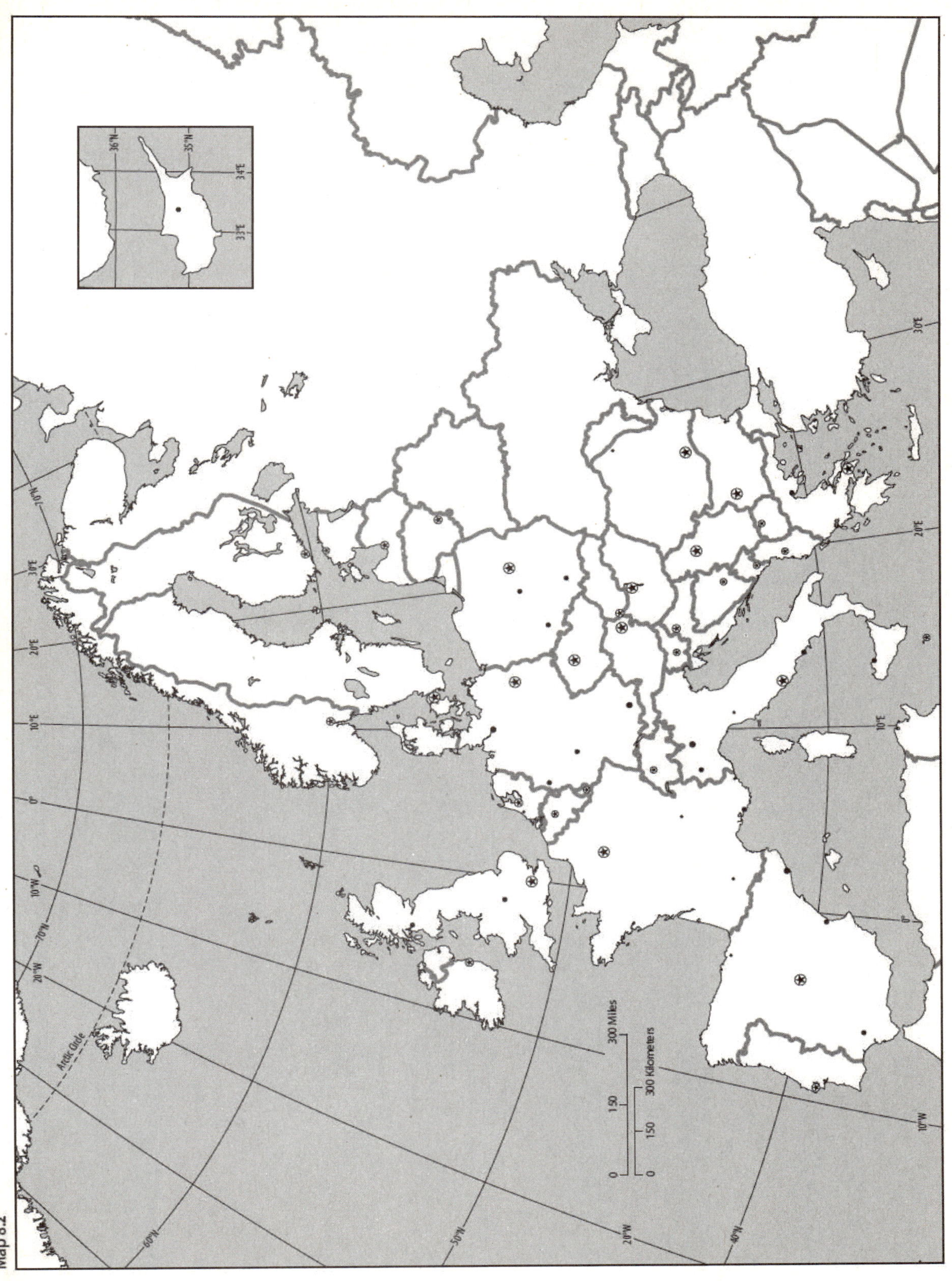

Map 8.2
Arctic Circle
0 150 300 Miles
0 150 300 Kilometers

Exercise One: Regional Patterns of Pollution and Industrialization

Using Figure 8.8 Environmental Issues in Europe (p. 226), Figure 8.31 Industrial Regions of Europe (p. 247), and Mapping Workbook Map 8.3, complete the following exercise.

Using Figure 8.8 "Environmental Issues in Europe" (p. 226) as a reference, use a colored pencil to shade in the European regions that have been affected by acid rain. Mark this shading or scheme on the map's legend. Using a different colored pencil, shade in the major European rivers that have been polluted. Mark this symbol on the map's legend. Once you have done this, answer the following questions.

1. Define acid rain.

____

____

2. What are the major causes of acid rain?

____

____

3. Describe the extent of acid rain damage in Europe. Specifically, what European countries have been greatly affected by acid rain?

____

____

____

____

4. Describe the extent of river pollution in Europe. Specifically, what European countries have polluted rivers?

____

____

____

____

Using Figure 8.31 "Industrial Regions of Europe" (p. 247) as a reference, use a red or black colored pencil (or a color that will contrast with the above shading scheme) to pattern in the older and newer European industrial areas (Use a different pattern for each). Mark these symbols on the map's legend. Once you have done this, answer the below questions.

5. Describe the location of the older European industrial areas. Specifically, in which European countries can they be found?

____

____

____

____

6. What type of manufacturing constitutes "older" industrial activities?

____

____

7. Does there appear to be a geographic pattern associated with these older industrial areas? If so, describe the pattern.

8. Describe the location of the newer European industrial areas. Specifically, in which European countries can they be found?

9. What type of manufacturing constitutes "newer" industrial activities?

10. Does there appear to be a geographic pattern associated with these newer industrial areas? If so, describe the pattern.

11. Compare the location of the European industrial areas and that of the extent of acid rain in the region. Is there a correlation between the locations of these polluted regions and that of industrial activity?

12. If you found a correlation, describe it. More specifically, do the polluted areas tend to correlate more with that of the older or the newer industrial activities?

13. Explain this correlation.

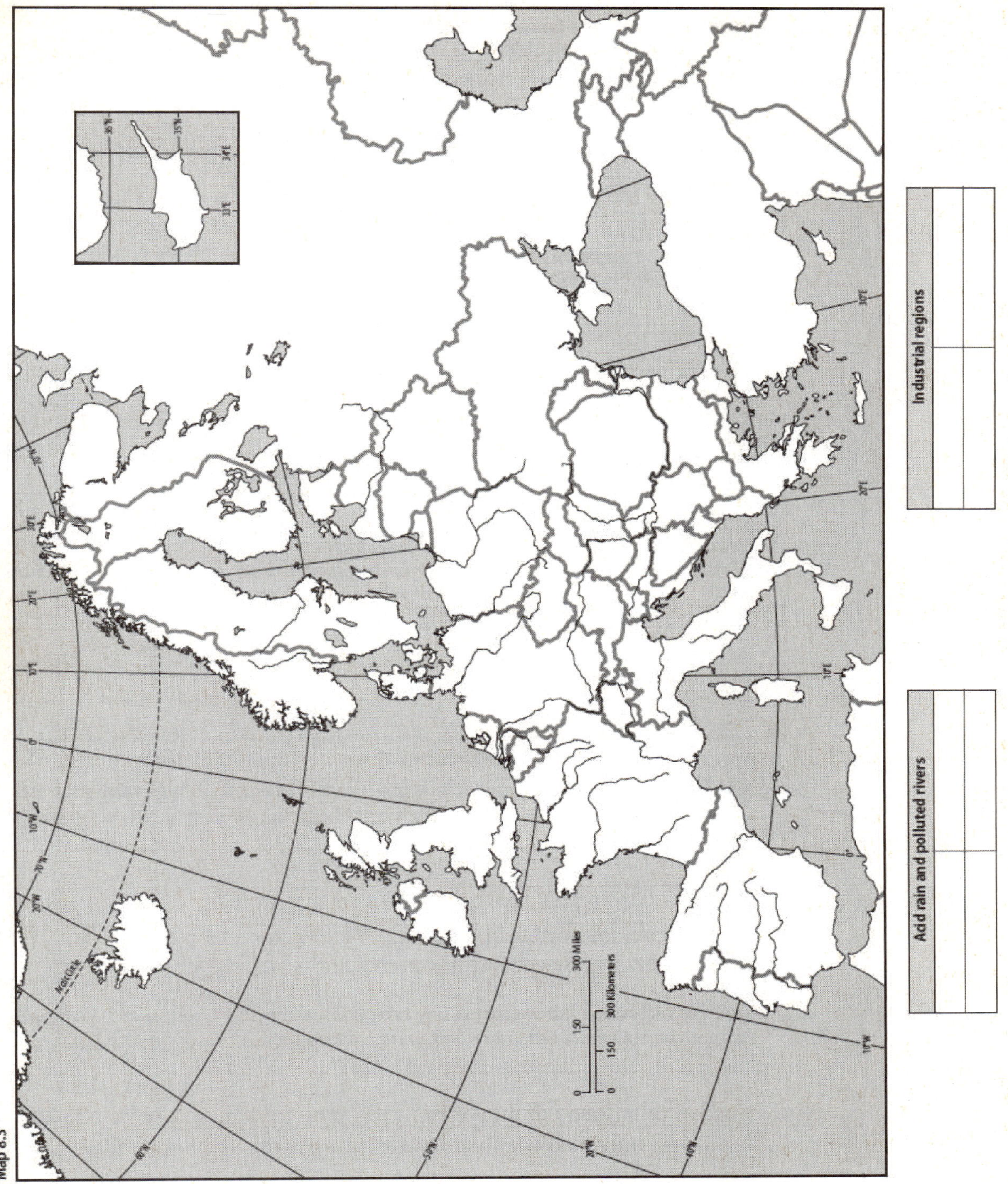

Industrial regions	

Acid rain and polluted rivers	

Exercise Two: European Language and Religion Patterns

Using Figure 8.20 Language Map of Europe (p. 235), Figure 8.22 Religions of Europe (p. 237), and Mapping Workbook Map 8.4, complete the following exercise.

Using Figure 8.20 "Language Map of Europe" (p. 235) as a reference, use a colored pencil to shade in the region where the Germanic language subfamily is dominant. Using a different color for each, shade in the other major language subfamilies in their respective locations on the map. Mark the appropriate shading schemes in the map's legend. Once you have done this, answer the following questions.

1. Describe the locations of each of the language subfamilies in Europe. More specifically, do you notice any overall patterns of clustering of language subfamilies in Europe (e.g., are all the Germanic languages clustered together)?

____

____

____

____

____

____

2. If so, what do you think accounts for these patterns? Explain.

____

____

____

____

Using Figure 8.22 "Religions of Europe" (p. 237) as a reference, use a red or black (or a color that will contrast with the above shading scheme) colored pencil to pattern in the region where the majority of the population is Roman Catholic. Using a different color for each, pattern in the other major religions in their respective locations on the map. Mark the appropriate patterning scheme in the map's legend. Once you have done this, answer the following questions.

3. Describe the locations of each of the modern religions in Europe. More specifically, do you notice any overall patterns of clustering of religions in Europe (e.g., are all the Protestants clustered together)?

____

____

____

____

____

____

4. If so, what do you think accounts for these patterns? Explain.

____

____

____

____

5. Examine your mapped patterns of language and religion. Do you notice any correlations between certain language subfamilies and modern religions practiced in European countries? If so, in which countries do these correlations exist?

6. If you answered yes to the above question, what do you think accounts for the correlations?

7. Which European countries possess non Indo–European language-speaking majorities?

8. Do you see a correlation between language and religion in the non Indo–European language speaking countries? If so, what are they and what do you think accounts for them? If not, why do you think they are different?

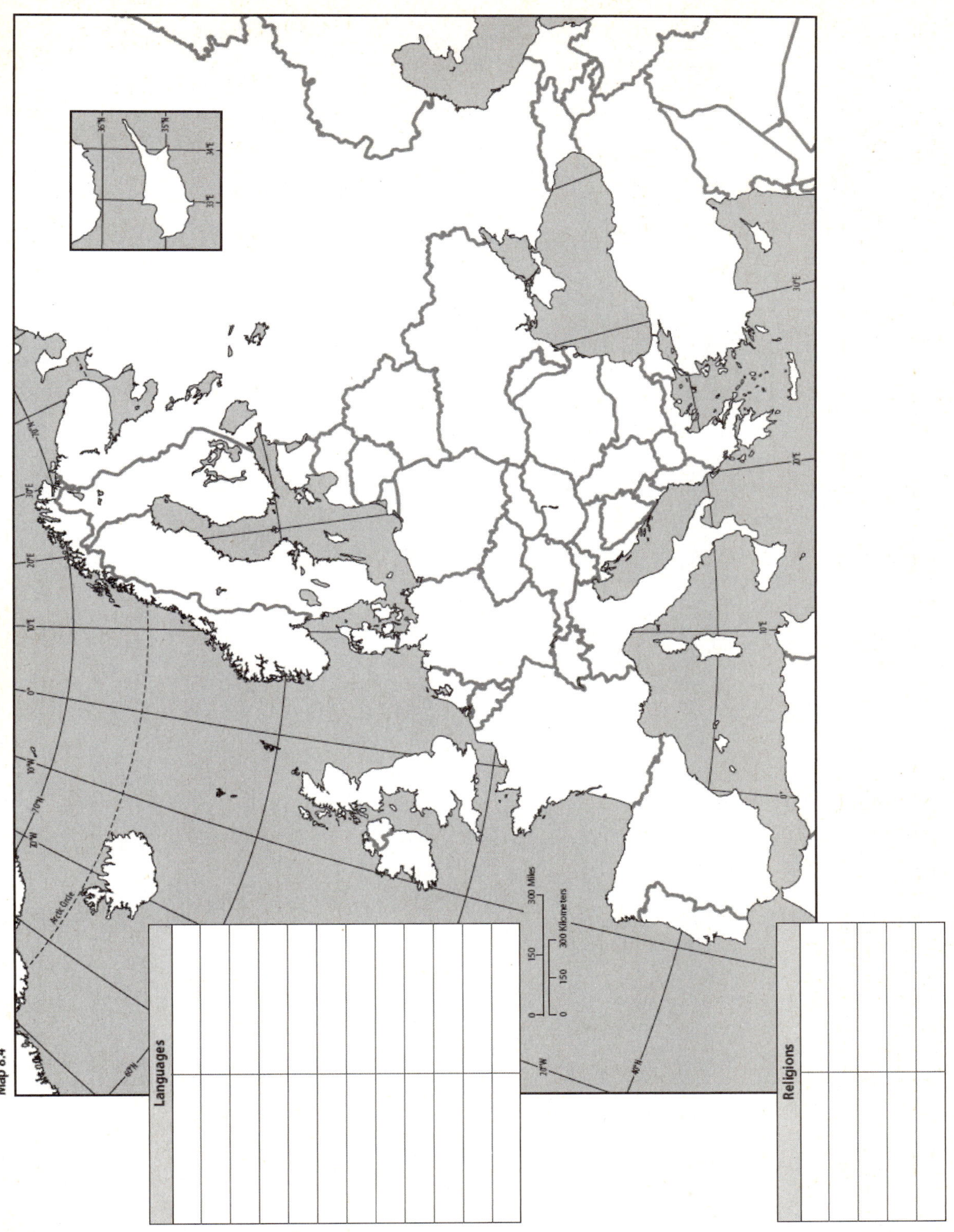
Map 8.4
Languages
Religions
300 Miles
300 Kilometers
Arctic Circle

Chapter Nine: The Russian Domain Mapping Workbook Exercises

Identify the following features on workbook Maps 9.1 and 9.2

Identify and label the following countries on Map 9.1

Belarus
Georgia
Moldova
Russia
Ukraine
Armenia

Identify and label the following cities on Map 9.1

Chelyabinsk
Dnepropetrovsk
Donetsk
Groznyy
Irkutsk
Kazan
Khabarovsk
Kharkov
Kiev
Krasnoyarsk
Minsk
Moscow
Norilsk
Novosibirsk
Odessa
Omsk
Samara
St. Petersburg
Tbilisi
Verkhoyansk
Vladivostok
Volgograd
Yakutsk
Yekaterinburg
Yerevan
Niznij Novgorod

Identify and label the following physical features on Map 9.2

Arctic Ocean
Baltic Sea
Barents Sea
Bering Sea
Black Sea
Caspian Sea
Caucasus Mountains
Dnieper River
Kamchatka Peninsula
Kola Peninsula
Kuril Islands
Lake Baikal
Lena River
Novaya Zemlya (Island)
Ob River
Sakhalin Island
Sea of Okhotsk
Siberia
Ural Mountains
Verkhoyansk Range
Volga River
Yenisey River
Yakutsk Basin

75°N
60°N
180°
45°N
30°N
165°E
15°N
0 400 800 Miles
0 400 800 Kilometers
150°E
135°E
120°E
105°E
90°E
75°E
60°E
165°W
180°
165°E
150°E
135°E
120°E
105°E
90°E
75°E
60°E
45°E
30°E
15°E
0°
75°N

Map 9.1

0 400 800 Miles
0 400 800 Kilometers
75°N
60°N
45°N
30°N
15°N
180°
165°E
165°W
150°E
135°E
120°E
105°E
90°E
75°E
60°E
45°E
30°E
15°E
0°

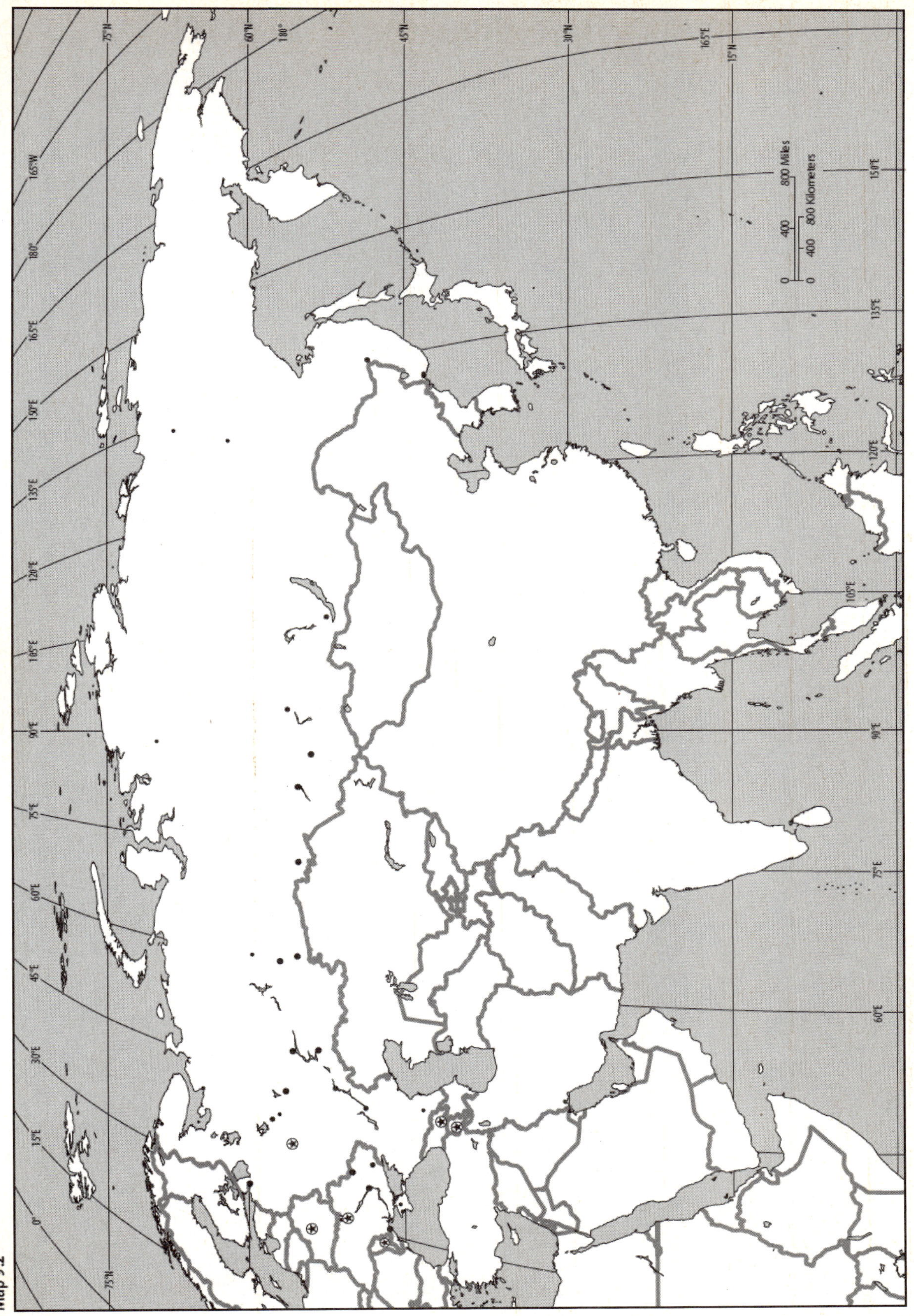

Map 9.2

Exercise One: Climate and Russian Domain Agricultural and Industrial Activities

Using Figure 9.7 Climate Map of the Russian Domain (p. 260), Figure 9.8 Agricultural Regions (p. 261), and Mapping Workbook Map 9.3, complete the following exercise.

Using Figure 9.7 "Climate Map of the Russian Domain" (p. 260) as a reference, use a colored pencil to shade in the regions that possess a Dfb climate. Using different colored pencils to represent each climate type, shade in the other majors climates within the region. Mark this shading scheme in the map's legend. Once you have done this, answer the following questions.

1. What are the major climates of the Russian Domain?

___

___

2. What are the general locations of these climates?

___

___

Using Figure 9.8 "Agricultural Regions" (p. 261) as a reference, use a red or black (or a color that contrasts with the shading scheme you created for climate) colored pencil to pattern in the major agricultural types within the region, using different patterns for diversified agriculture, large-scale grain production, urban truck farming, and humid subtropical specialized agricultural production. Also, use a different pattern to indicate the locations of tundra, taiga, drylands, and mountains. Mark this patterning scheme in the map's legend. Once you have done this, answer the following questions.

3. What is (are) the location(s) of diversified agriculture within the Russian Domain?

___

___

4. What is (are) the location(s) of large-scale grain production within the Russian Domain?

___

___

5. What is (are) the location(s) of urban truck farming within the Russian Domain?

___

___

6. What is (are) the location(s) of humid subtropical specialized agriculture within the Russian Domain?

___

___

7. How well do these agricultural activities correlate with the biomes and landforms (i.e., tundra, taiga, drylands, and mountains) in the region?

___

___

___

___

8. Examine the agricultural regions and their location relative to the climate types, biomes and landforms you shaded in on the map. Which biomes, landforms and climate does diversified agriculture correlate with, if any?

9. How would you explain the correlation between this agricultural activity and the biomes, landforms, and climate within the region where it prevails? If there was no correlation, how would you explain the lack of one?

10. Which biomes, landforms and climate correlate with large-scale grain production, if any?

11. How would you explain the correlation between this agricultural activity and the biomes, landforms, and climate within the region where it prevails? If there was no correlation, how would you explain the lack of one?

12. Which biomes, landforms and climate correlate with urban truck farming, if any?

13. How would you explain the correlation between this agricultural activity and the biomes, landforms, and climate within the region where it prevails? If there was no correlation, how would you explain the lack of one?

14. Which biomes, landforms and climate correlate with humid subtropical specialized agricultural production, if any?

15. How would you explain the correlation between this agricultural activity and the biomes, landforms, and climate within the region where it prevails? If there was no correlation, how would you explain the lack of one?

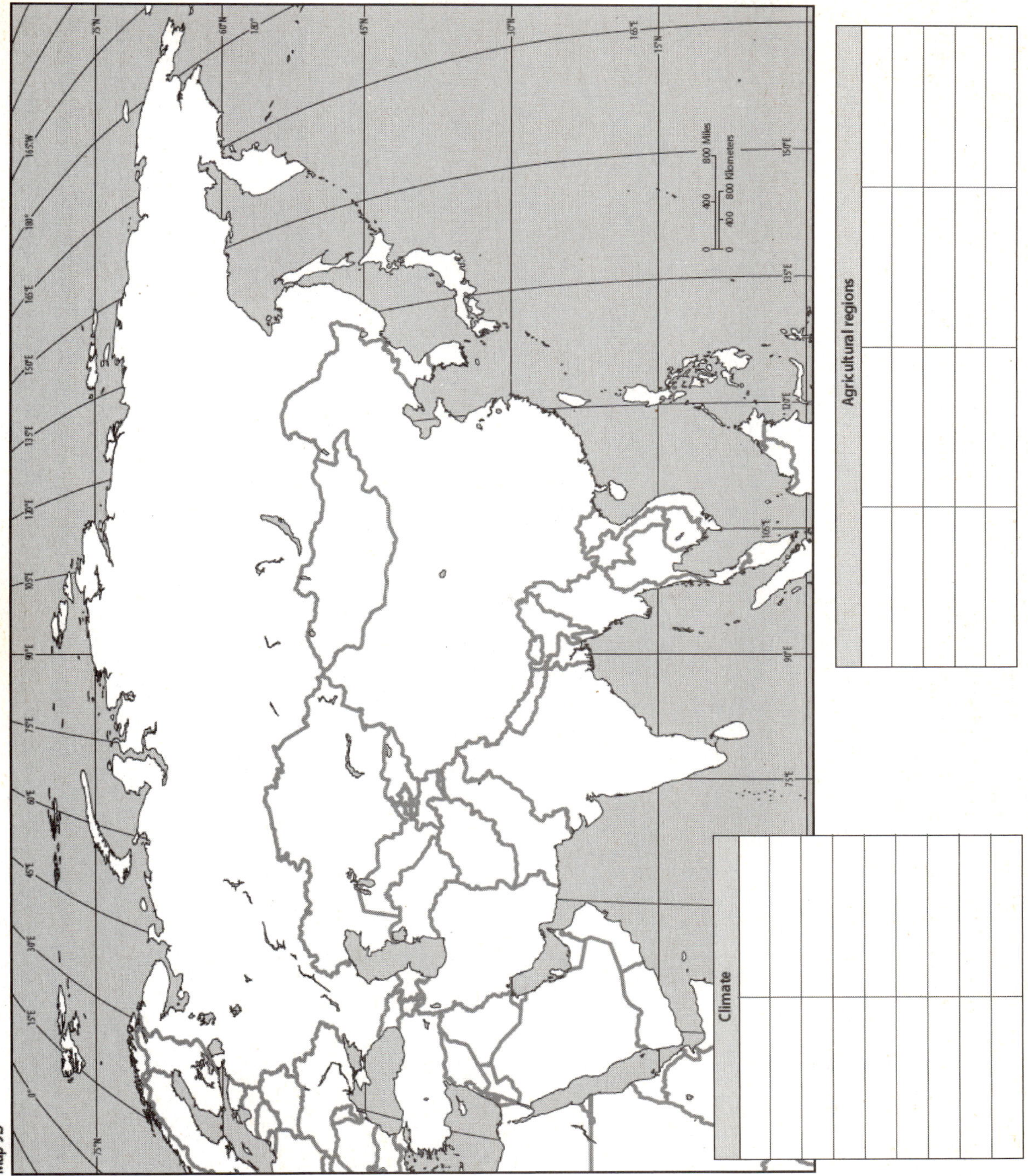
Map 9.3
0 400 800 Miles
0 400 800 Kilometers
Climate
Agricultural regions

Exercise Two: Regional Patterns of Pollution and Industrialization in the Russian Domain

Using Figure 9.4 Environmental Issues in the Russian Domain (p. 258), Figure 9.32 Major Natural Resources and Industrial Zones (p. 279), and Mapping Workbook Map 9.4, complete the following exercise.

Using Figure 9.32 "Major Natural Resources and Industrial Zones" (p. 279) as a reference, use a colored pencil to shade in the major regions of forestry in the Russian Domain. Mark this shading scheme on the map's legend. Use a different colored pencil to shade in the major regions of manufacturing within the Russian Domain. Create symbols for the major natural resources found in the Russian Domain and label these on the map in their respective locations. Use a different colored pencil for each symbol. Mark these symbols on the map's legend. Once you have done this, answer the following questions.

1. Describe the location of the natural resources found in the Russian Domain. Specifically, in which regions are they found?

2. Where is the greatest concentration of natural resources in the Russian Domain?

Now examine the mapped location of the various industrial activities in the Russian Domain.

3. Do you see a correlation between the location of the Russian Domain's natural resources and the location of the Russian Domain's major industrial regions? If so, what is it?

4. Are there any areas where there are natural resources and yet there are few industrial activities? If so, where?

5. If you answered yes to the above question, what could explain this phenomenon?

Using Figure 9.4 "Environmental Issues in the Russian Domain" (p. 258) as a reference, use a colored pencil to pattern in the regions of the Russian Domain that have been affected by acid rain. Mark this patterning scheme on the map's legend. Now, using a different colored pencil, create different patterns and indicate the regions of forest damage, radioactive contamination, coastal pollution, salinization, and polluted rivers. Mark these patterns in the map's legend. Once you have done this, answer the following questions.

6. Compare the locations of the Russian Domain industrial activities and that of the extent of acid rain in the region. Is there a correlation between the locations of these polluted regions and that of industrial activity?

---

7. If you found a correlation, describe and explain it.

---

8. Compare the locations of the Russian Domain's industrial activities and that of the extent of forest damage in the region. Is there a correlation between the locations of these polluted regions and that of industrial activity?

---

9. If you found a correlation, describe and explain it.

---

10. Compare the locations of the Russian Domain's industrial activities and that of the extent of radioactive contamination in the region. Is there a correlation between the locations of these polluted regions and that of industrial activity?

---

11. If you found a correlation, describe and explain it.

---

12. Compare the locations of the Russian Domain's industrial activities and that of the extent of coastal pollution in the region. Is there a correlation between the locations of these polluted regions and that of industrial activity?

---

13. If you found a correlation, describe and explain it.

14. Compare the locations of the Russian Domain's industrial activities and that of the extent of salinization in the region. Is there a correlation between the locations of these polluted regions and that of industrial activity?

15. If you found a correlation, describe and explain it.

16. Compare the locations of the Russian Domain's industrial activities and that of the extent of river pollution in the region. Is there a correlation between the locations of these polluted regions and that of industrial activity?

17. If you found a correlation, describe and explain it.

Map 9.4

Environmental Issues
Natural resources and Industrial zones
0 400 800 Miles
0 400 800 Kilometers

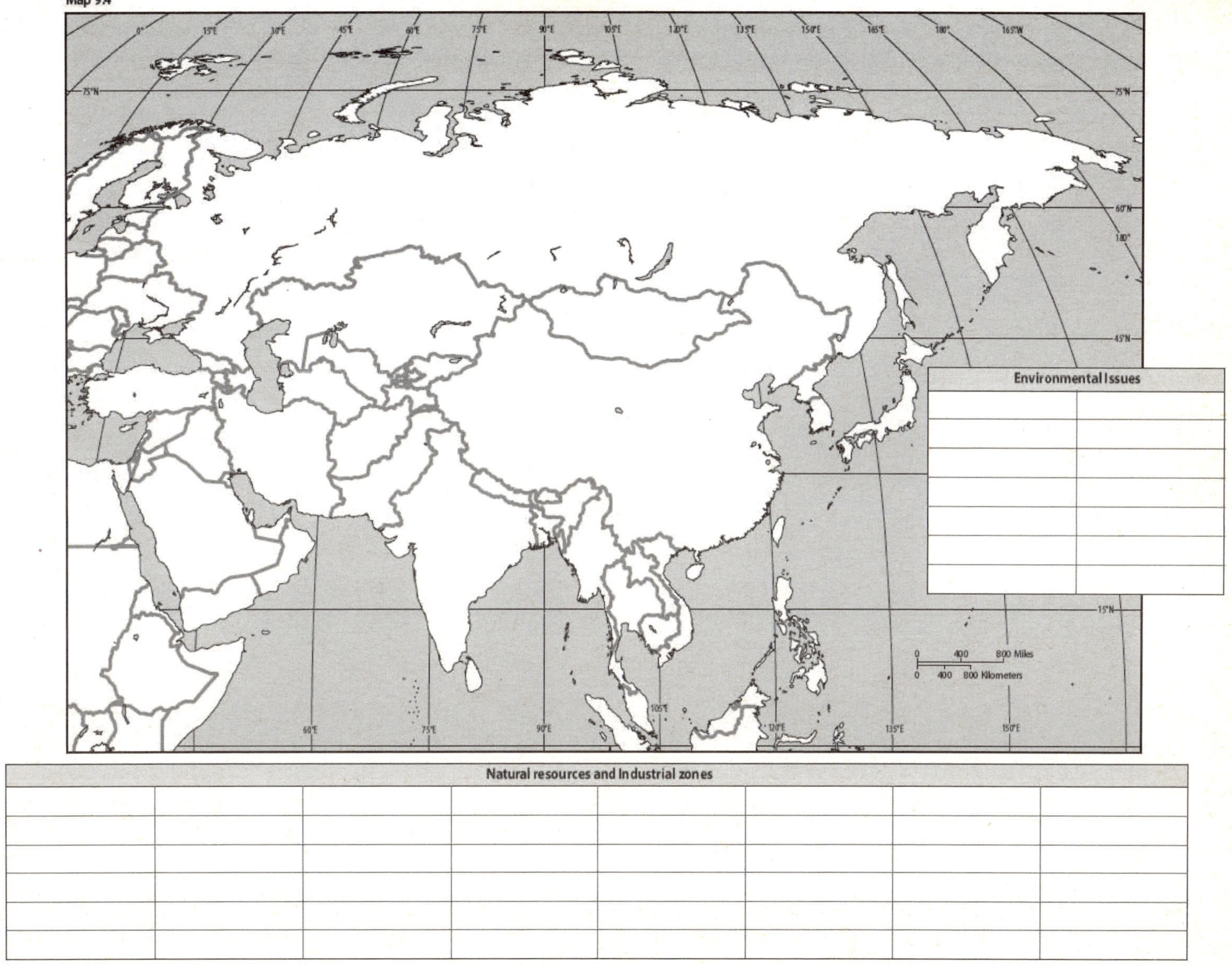

Chapter Ten: Central Asia Mapping Workbook Exercises

Identify the following features on workbook Maps 10.1 and 10.2

Identify and label the following countries on Map 10.1

Afghanistan
Kazakhstan
Kyrgyzstan
Mongolia
Tajikistan
Turkmenistan
Uzbekistan

Identify and label the following cities on Map 10.1

Almaty
Aqtau
Ashgabat
Astana
Baotou
Bishkek
Dushanbe
Hohhot
Kabul
Kandahar
Lhasa
Nukus
Oral
Shymkent
Tashkent
Ulaanbaatar
Urumqi

Identify and label the following physical features on Map 10.2

Altai Mountains
Amu Darya River
Aral Sea
Caspian Sea
Gobi Desert
Helmand River
Hindu Kush
Ili River
Kara -Bogaz Gol
Kara Kum Canal
Kara Kum Desert
Kyzyl Kum Desert
Lake Balqash
Pamir Mountains
Syr Darya River
Taklamakan Desert
Tien Shan
Turfan Depression
Ural River

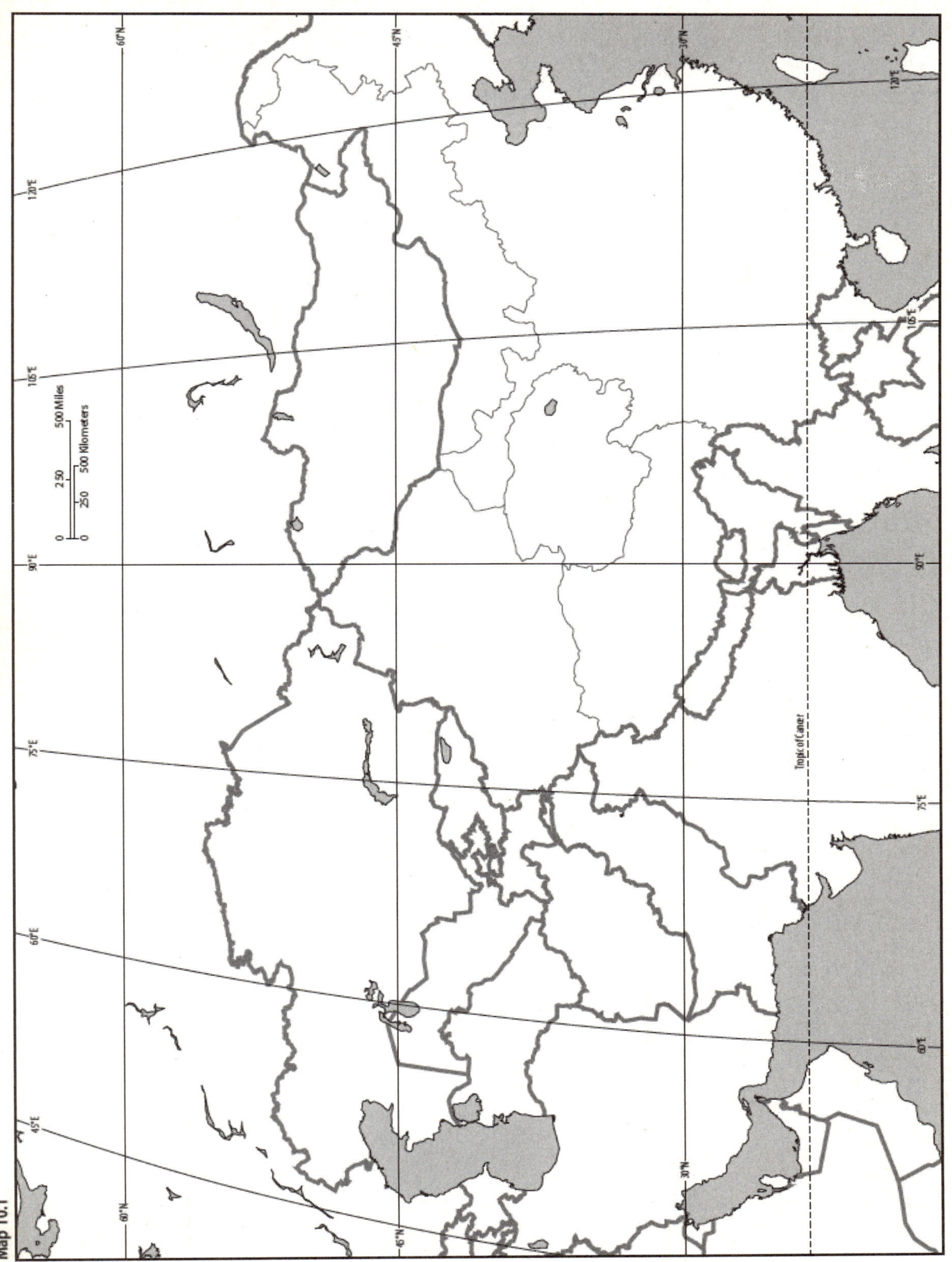
Map 10.1
0 250 500 Miles
0 250 500 Kilometers
120°E
105°E
90°E
75°E
60°E
45°E
60°N
45°N
30°N
Tropic of Cancer

60°N
45°N
30°N
120°E
105°E
90°E
75°E
60°E
45°E
0 250 500 Miles
0 250 500 Kilometers
Tropic of Cancer

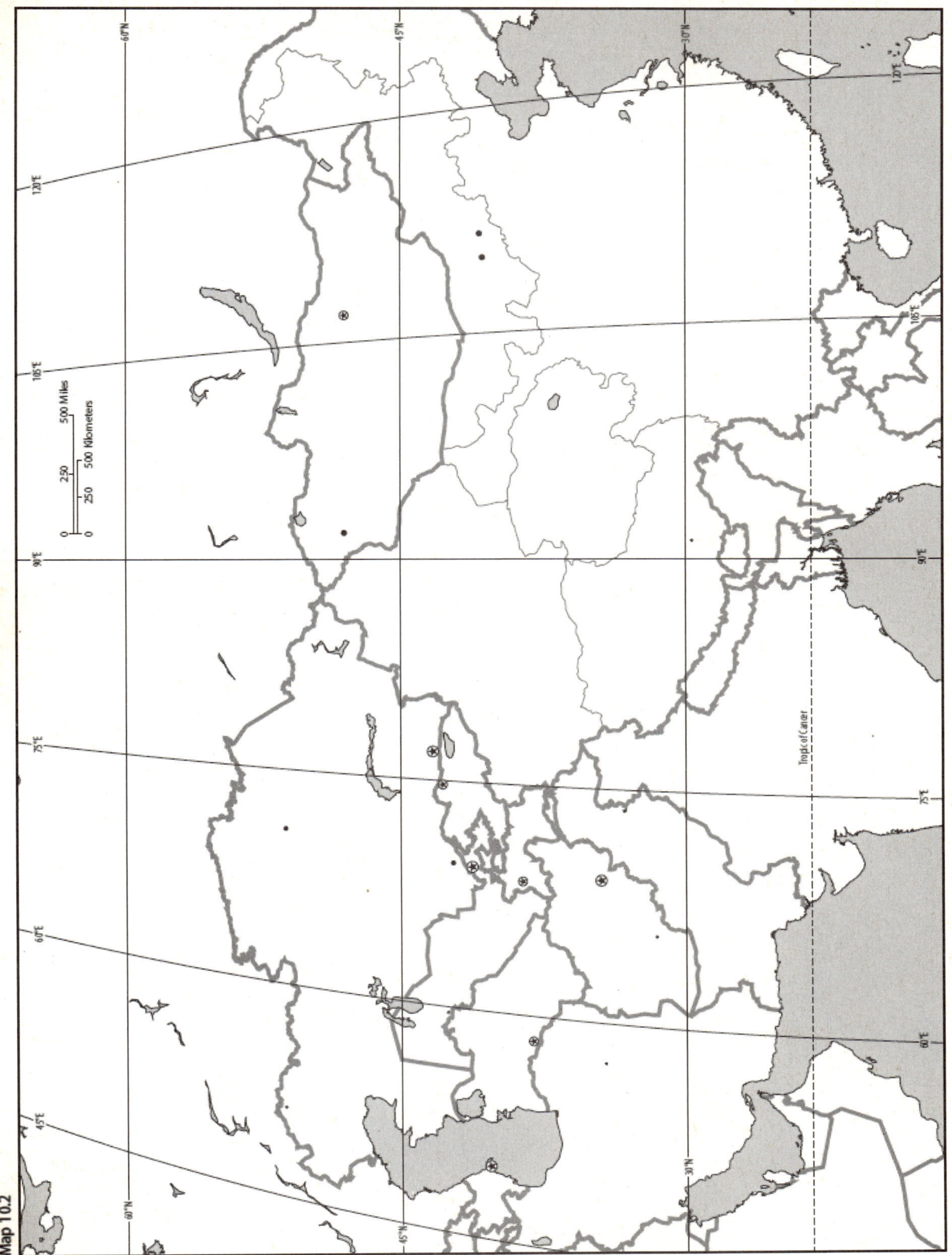

Map 10.2

Exercise One: Landforms, Climate, and Settlement in Central Asia

Using Figure 10.1 Central Asia (physical features) (pp. 286-287), Figure 10.6 Climates of Central Asia (p. 290), Figure 10.8 Population Density in Central Asia (p. 292), Figure 10.9 Population Patterns in Xinjiang's Tarim Basin (p. 293), and Mapping Workbook Map 10.3, complete the following exercise.

Using Figure 10.1 "Central Asia" (physical features) (pp. 286-287) as a reference, use a colored pencil to shade in the major Central Asian desert regions. Using different colored pencils to represent each landform, shade in the Central Asian highlands, steppes, and plateaus in their appropriate locations. Use a black or red (or another color that will contrast with the above shading scheme) colored pencil and draw lines indicating the approximate location of the rivers in the region. Mark the shades and symbols for the landforms and rivers in the map's legend. Once you have done this, answer the following questions.

1. Describe the location of the deserts in Central Asia.

2. What general criteria are used to classify a region as a deserts?

3. What is a steppe?

4. Describe the location of the steppes in Central Asia.

5. Describe the location of the highlands in Central Asia.

6. How do the locations of rivers generally correspond to the location of highlands in the region?

Using Figure 10.6 "Climates of Central Asia" (p. 290) as a reference, use colored pencils to pattern in their respective locations the major Central Asian climate types. Use a different color to represent each climate type. Mark the patterns for the climates in the map's legend. Once you have done this, answer the following questions.

7. Describe the general climatic pattern in Central Asia. What is the predominant climate in the region?

8. Examine the pattern of Central Asian climate and the pattern of Central Asian landforms together. Are there correlations between the location of the various landforms and the major climates in the region? If so, describe what they are.

9. How would the landforms within Central Asia have an effect (i.e., control) on the dominant climate in the region?

10. How might the landforms also cause subtle variations in the Central Asian climate patterns?

Using Figure 10.8 "Population Density in Central Asia" (p. 292) as a reference, use a red or black (or another color that contrasts with the shading and patterning schemes you have already established) colored pencil and color in the Central Asian regions possessing high concentrations of population. Mark the shades or pattern for population concentration in the map's legend. Once you have done this, answer the following questions.

11. Describe the regions of highest population concentration.

12. Describe the regions of lowest population concentration.

Examine the regions of the highest and lowest population concentration and how this concentration corresponds to the locations of the various landforms and climates you placed on your map.

13. How does Central Asian population correlate with landforms? Explain this correlation.

14. How does Central Asian population correlate with climate? Explain this correlation.

Look at Figure 10.9 "Population Patterns in Xinjiang's Tarim Basin" (p. 293), closely paying attention to the pattern of population settlement and landforms.

15. Does the settlement within the Tarim Basin reflect the wider patterns of Central Asian population settlement?

16. If so, explain why you believe this is the case. If not, explain why you think population settlement within the Tarim Basin differs from the rest of the region.

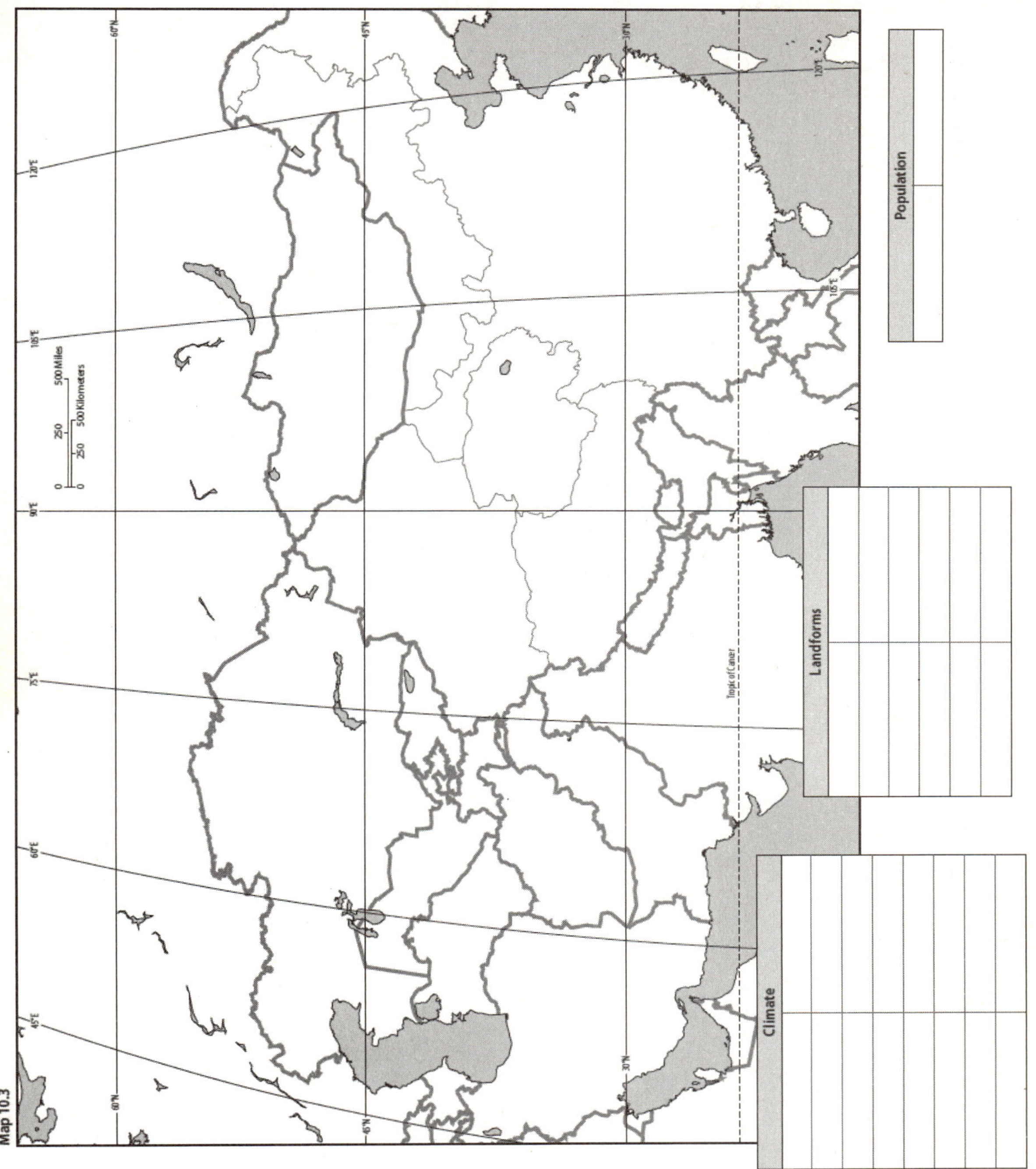

Map 10.3
Population
Landforms
Climate
0 250 500 Miles
0 250 500 Kilometers
Tropic of Cancer

Exercise Two: Valley-ism in Afghanistan

Using Figure 10.1 Central Asia (physical features) (pp. 286-287), Figure 10.16 Afghanistan's Ethnic Patchwork (p. 297), Figure 10.8 Population Density in Central Asia (p. 292), and Mapping Workbook Map 10.4, complete the following exercise.

Using Figure 10.16 "Afghanistan's Ethnic Patchwork" (p. 297) as a reference, use a colored pencil to shade in the regions that possess the majority of the Tajik speakers. Using a different color to represent each language, indicate the location of these other languages on your map by shading them in their respective regions. Mark the shades in the map's legend. Once you have done this, answer the following.

Describe the general patterns of each of the major languages within Central Asia.

1. Tajik ____

____

2. Hazara ____

____

3. Aimak ____

____

4. Pashtun ____

____

5. Nuristani ____

____

6. Pashai ____

____

7. Baluchi ____

____

8. Uzbek ____

____

9. Turkmen ____

____

10. Kyrgyz ____

____

11. Brahui ____

____

12. Which of these languages display the greatest amount of overlap?

13. In the regions where this overlap occurs, how different are these languages from each other?

14. How might the linguistic differences cause cultural and political tensions in these areas?

Using Figure 10.1 "Central Asia" (physical features) (pp. 286-287) as a reference, use a red or black (or another color that would contrast with the shading scheme you established) colored pencil to pattern in the major Afghanistan landforms, using different colored pencils and patterns to represent the various landforms that are present in Afghanistan. Mark the patterns in the map's legend. Once you have done this, answer the following questions.

15. What is the predominant landform in Afghanistan?

16. How extensive is this landform in Afghanistan?

17. Describe the pattern of mountains in Afghanistan. Which way do the mountains trend (e.g., north-south, east-west, etc.)?

Using Figure 10.8 "Population Density in Central Asia" (p. 292) as a reference, use a red or black (or another color that would contrast with the shading scheme you established) colored pencil to pattern in the Afghanistan population distribution. Mark the pattern in the map's legend. Once you have done this, answer the following questions.

18. Examine the physical geography of Afghanistan and compare it with the pattern of population settlement in the country. How does settlement correlate with Afghanistan's physical geography?

19. How do you explain this correlation?

20. Examine the physical regions, the patterns of population settlement, and patterns of language you drew on your map. How do the patterns of population settlement correlate with the patterns of language in the country?

____

____

____

____

21. How do you believe these patterns affect the cultural relations, political cohesion and stability within the country?

____

____

____

____

Map 10.4

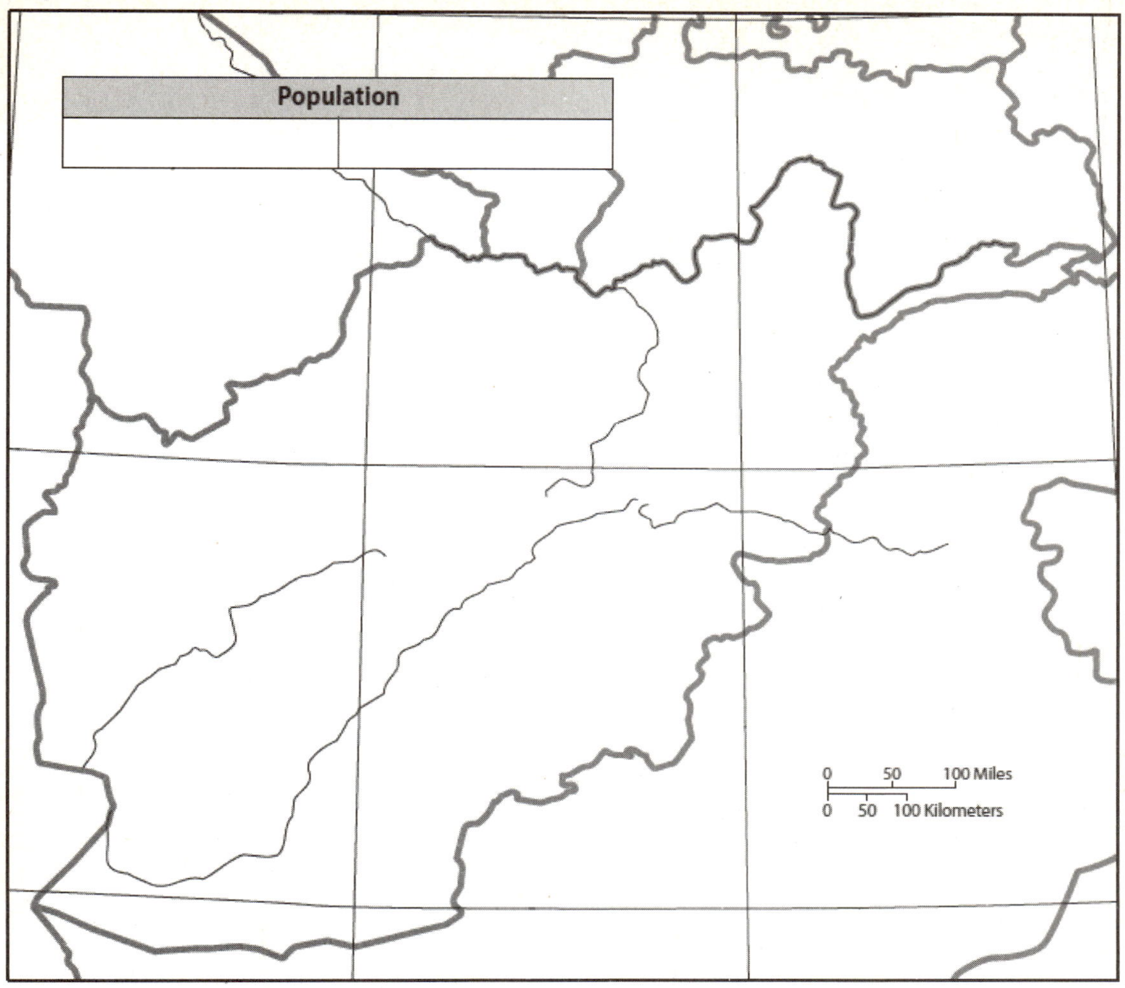

Languages	

Landforms	

Chapter Eleven: East Asia Mapping Workbook Exercises

Identify the following features on workbook Maps 11.1 and 11.2

Identify and label the following countries on Map 11.1

China
Japan
North Korea
South Korea
Taiwan

Identify and label the following cities on Map 11.1

Beijing
Changchun
Chengdu
Chongqing
Guangzhou
Hangzhou
Harbin
Hiroshima
Hong Kong
Kobe
Kwangju
Kyoto
Lhasa
Macao
Nanjing
Osaka
Pusan
Pyongyang
Sapporo
Seoul
Shanghai
Shenyang
Taipei
Tangshan
Tianjin
Tokyo
Urumqi
Wuhan
Xi'an

Identify and label the following physical features on Map 11.2

Amur River
Bay of Bengal
Cheju (Island)
East China Sea
Gobi Desert
Grand Canal
Gulf of Tonkin
Hainan (Island)
Himalayas
Hokkaido
Hong (Red) River
Honshu
Huang He
Kyushu
Mekong River
Pacific Ocean
Plateau of Tibet
Ryukyu Islands
Sea of Japan
Shikoku
South China Sea
Taklamakan Desert
Ussuri River
Yangtze River
Yellow Sea

Map 11.1

Map 11.2

Exercise One: Populated and Unpopulated Regions of China

Using Figure 11.1 East Asia (physical landscape) (pp. 310-311), Figure 11.2 Environmental Issues in East Asia (p. 312), Figure 11.15 Population Map of East Asia (p. 320), and Mapping Workbook Map 11.3, complete the following exercise.

Using Figure 11.1 "East Asia" (physical landscape) (pp. 310–311) as a reference, use a colored pencil to shade in the low elevation regions of China (0–500 feet above sea level). Using a different colored pencil, shade in the higher elevation regions of China (500–2000 feet above sea level). Using a pencil that is a different color from the first two you used, shade in the highest elevation regions of China (4000 + feet above sea level). Mark the shading scheme in the map's legend. Once you have done this, answer the following questions.

1. Describe the location of the low elevation regions in China.

____

____

____

2. Describe the location of the higher elevation regions in China.

____

____

____

3. Describe the location of the highest elevation regions in China.

____

____

____

Using Figure 11.15 "Population Map of East Asia" (p. 320) as a reference, use a red or black (or a color that will contrast with the elevation shading pattern you established) colored pencil to pattern the locations of Chinese population. Mark the patterning scheme in the map's legend. Once you have done this, answer the following questions.

4. What regions in China possess the highest population densities?

____

____

5. What do you think accounts for the clustering of people into these regions?

____

____

6. Compare the patterns of elevation and population settlement on your map. Do physical features seem to play a role in high population densities? If so, what are the physical characteristics of the landscape that appear to promote population?

____

____

____

7. Where are the Asian regions that possess the lowest population densities?

____

____

8. What do you think accounts for the clustering of people in these regions?

9. Do physical features seem to play a role in low population densities? If so, what are the physical characteristics of the landscape that appear to limit population?

Using Figure 11.2 "Environmental Issues in East Asia" (p. 312) as a reference, use a colored pencil to pattern the location of forest areas in China. Using a different colored pencil for each, pattern in the Chinese regions of extensive deforestation, desertification, severe soil erosion, coastal pollution, and high risk of flooding. Mark the patterning scheme in the map's legend. Once you have done this, answer the following questions.

10. Where are the greatest locations of extensive deforestation in China?

11. Where are the greatest locations of desertification in China?

12. Where are the greatest locations of severe soil erosion in China?

13. Where are the greatest locations of coastal pollution in China?

14. Examine the location of the above environmental hazards and how they correspond to population within China. Does there appear to be a correlation between the regions of highest population density and the location of these environmental hazards? If so, specifically describe these correlations.

15. How are these environmental impacts related to population? Elaborate.

16. Examine the location of the regions of China subject to the greatest risk of flooding and how it corresponds to population within China. Where are the regions of greatest flooding risk?

17. Does there appear to be a correlation between the regions of highest population density and the location of flooding risk? If so, specifically describe this correlation.

18. How will flooding have an impact on the Chinese population in the regions where it is most prevalent?

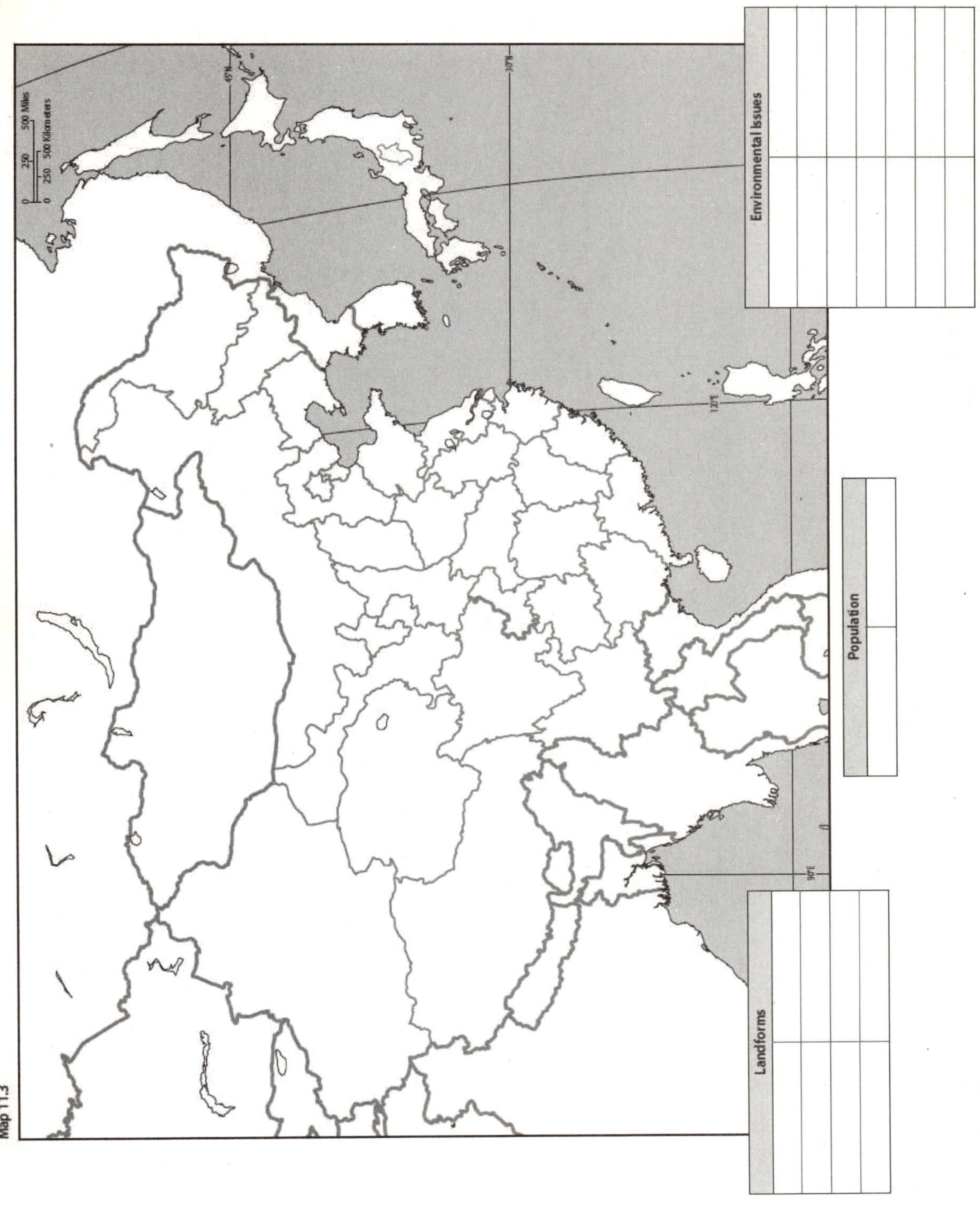
Map 11.3
0 250 500 Miles
0 250 500 Kilometers
Environmental Issues
Population
Landforms

Exercise Two: Japanese Physical Geography and Settlement

Using Figure 11.9 Japan's Physical Geography (p. 316), Figure 11.15 Population Map of East Asia (p. 320), Figure 11.19 Urban Concentration in Japan (p. 323), and Mapping Workbook Map 11.4, complete the following exercise.

Using Figure 11.9 "Japan's Physical Geography" (p. 316) as a reference, use a colored pencil to shade in the approximate locations of hill lands and mountains in Japan. Using a different color pencil shade in the approximate locations of diluvial plains and lowlands of new alluvium in Japan. Using a different colored pencil, draw lines representing the approximate locations of plate boundaries. Label the plates and plate boundaries. Using a different colored pencil, draw lines indicating the regions of tsunami activity within Japan. Using different colored pencils, create point symbols and indicate the location for earthquake epicenters and major volcanic eruptions. Mark the shading scheme and symbols on the map's legend. Once you have done this, answer the following questions.

1. How many tectonic plates converge on Japan and what are their names?

____

____

2. Does the location of these plates correspond to the location of volcanic and earthquake activity? If so, what may account for it?

____

____

____

____

Using Figure 11.15 "Population Map of East Asia" (p. 320) as a reference, use a red or black (or a different color that will contrast with the shading scheme you created) colored pencil to indicate the Japanese regions of highest population density. Mark the population shading scheme on the map's legend. Once you have done this, answer the following questions.

3. Where are the areas of highest population density in Japan?

____

____

4. Where are the areas of lowest population density in Japan?

____

____

5. Examine the patterns of population and physical features you shaded on your map. Do you notice a correlation between the location of physical features and population concentration? If so, what are they?

____

____

____

____

6. What accounts for the patterns of population settlement in Japan?

____

____

____

____

Using Figure 11.19 "Urban Concentration in Japan" (p. 323) as a reference, use a red or black (or a different color that will contrast with the shading scheme you created) colored pencil to indicate the Japanese region's urban concentration. Mark the population shading scheme on the map's legend. Once you have done this, answer the following questions.

7. How does the location of this major urbanized Japanese region correlate to the physical features you labeled on your map?

8. How has the physical geography had an impact on this high concentration of Japanese urban settlement? Explain.

9. How does the pattern of Japanese urban concentration correlate with the locations of earthquake epicenters?

10. How does the pattern of Japanese urban concentration correlate with the locations of major volcanic eruptions?

11. How does the pattern of Japanese urban concentration correlate with the locations of greatest potential tsunami activity?

12. How might any of these hazards (i.e., earthquakes, volcanic eruptions, and tsunamis) would affect the Japanese population?

Map 11.4

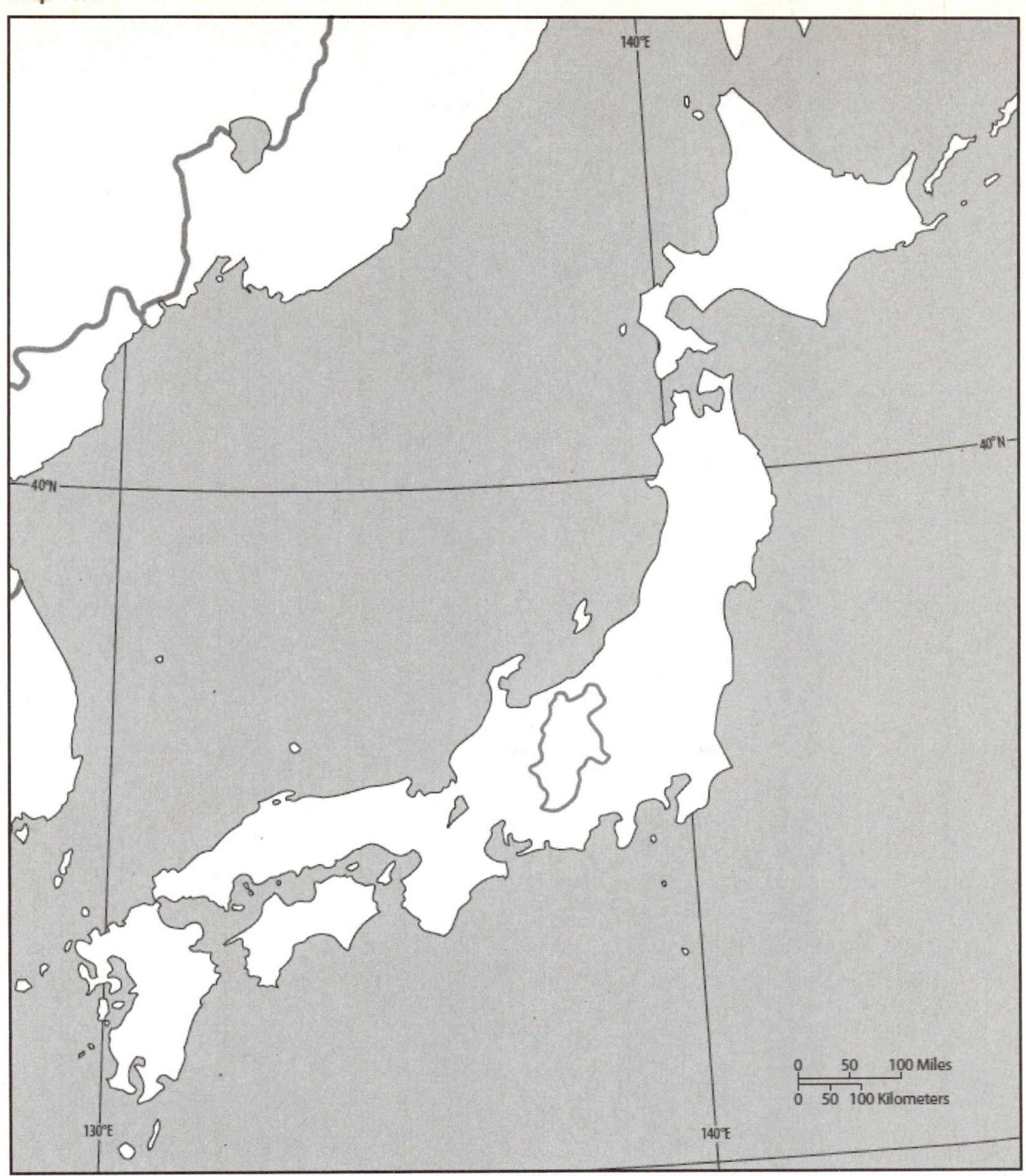

Landforms	

Population	

Urban concentration	

Chapter Twelve: South Asia Mapping Workbook Exercises

Identify the following features on workbook Maps 12.1 and 12.2

Identify and label the following countries on Map 12.1

Bangladesh
Bhutan
India
Maldives
Nepal
Pakistan
Sri Lanka

Identify and label the following cities on Map 12.1

Agra
Ahmadabad
Amritsar
Bengaluru (Bangalore)
Chennai (Madras)
Chittagong
Colombo
Delhi
Dhaka
Gwadar
Hyderabad (India)
Hyderabad (Pakistan)
Islamabad
Jaffna
Jaipur
Kanpur
Karachi
Kathmandu
Kolkata (Calcutta)
Lahore
Male
Multan
Mumbai (Bombay)
New Delhi
Surat
Thimphu

Identify and label the following physical features on Map 12.2

Andaman Islands
Andaman Sea
Arabian Sea
Aravalli Range
Bay of Bengal
Bhima River
Brahmaputra River
Cape Comorin
Central Makran Range
Coromandel Coast
Deccan Plateau
Eastern Ghats
Gaghara River
Ganges Delta
Ganges Plain
Ganges River
Godavari River
Godwin Austen Peak (K2)
Gulf of Khambhat
Gulf of Kutch
Indian Ocean
Indus River
Jhelum River
Karakoram Range
Kashmir
Kathiawar Peninsula
Krishna River
Lakshadweep
Malabar Coast
Mt. Everest
Narmada River
Nicobar Islands
Palk Strait
Rann of Kutch
Ravi River
Satpura Range
Sulaiman Range
Sundarbans
Sutlej River
Thar Desert
Vindhya Range
Western Ghats
Yamuna River

Map 12.1
0 100 200 Miles
0 100 200 Kilometers
20°N
10°N
100°E
90°E
80°E
70°E

Map 12.2
0 100 200 Miles
0 100 200 Kilometers
20°N
10°N
100°E
90°E
80°E
70°E

Exercise One: South Asian Patterns of Religion and Territorial Tension

Using Figure 12.17 Religious Geography of South Asia (p. 356), Figure 12.25 Geopolitical Issues in South Asia (p. 361), Figure 12.26 Geopolitical Change (p. 362), and Mapping Workbook Map 12.3, complete the following exercise.

Using Figure 12.17 "Religious Geography of South Asia" (p. 356) as a reference, use a colored pencil to shade in the major regions of Hinduism in South Asia. Using a different colored pencil for each religion, indicate the regions where the other major religions are prevalent in South Asia. Mark your shading scheme in the map's legend. Once you have done this, answer the following questions.

1. In which South Asian region(s) do you find the majority of Hindus?

____

____

2. In which South Asian region(s) do you find the majority of Muslims?

____

____

3. In which South Asian region(s) do you find the majority of Buddhists?

____

____

4. In which South Asian region(s) do you find the majority of Sikhs?

____

____

5. In which South Asian region(s) do you find the majority of Christians?

____

____

6. In which South Asian region(s) do you find the majority of Jains?

____

____

7. In which South Asian region(s) do you find the majority of adherents to tribal religions?

____

____

8. Do the adherents to these religions appear to be clustered within their own groups, or are they rather more evenly mixed throughout the region? Elaborate.

____

____

____

____

Using Figure 12.25 "Geopolitical Issues in South Asia" (p. 361) as a reference, use a colored pencil to pattern in the areas claimed by India, controlled by China. Using a different colored pencil and pattern to represent each conflict, indicate the locations of these conflicts on the South Asia map. Mark your patterning scheme in the map's legend. Once you have done this, answer the following questions.

9. Describe the various territorial conflicts in South Asia. Be sure to note their locations.

____

____

____

____

10. On the map you created, examine the pattern of religions and territorial conflict together. Do you see a correlation between any of the locations of the major South Asian religions and the locations of territorial conflict?

____

____

11. If so, what and where are they? Describe the nature of the conflicts.

____

____

____

____

____

____

In your textbook, examine Figure 12.26 "Geopolitical Change" (p. 362) and study the evolution of the modern South Asian national borders.

12. How might the progress of South Asia from a unified empire in 1700 to the collection of countries that it possesses today have generated the contemporary landscape of territorial conflicts? Elaborate.

____

____

____

____

____

____

____

____

13. Are these conflicts merely a matter of territory, or does religion play a role? Elaborate.

____

____

____

____

____

____

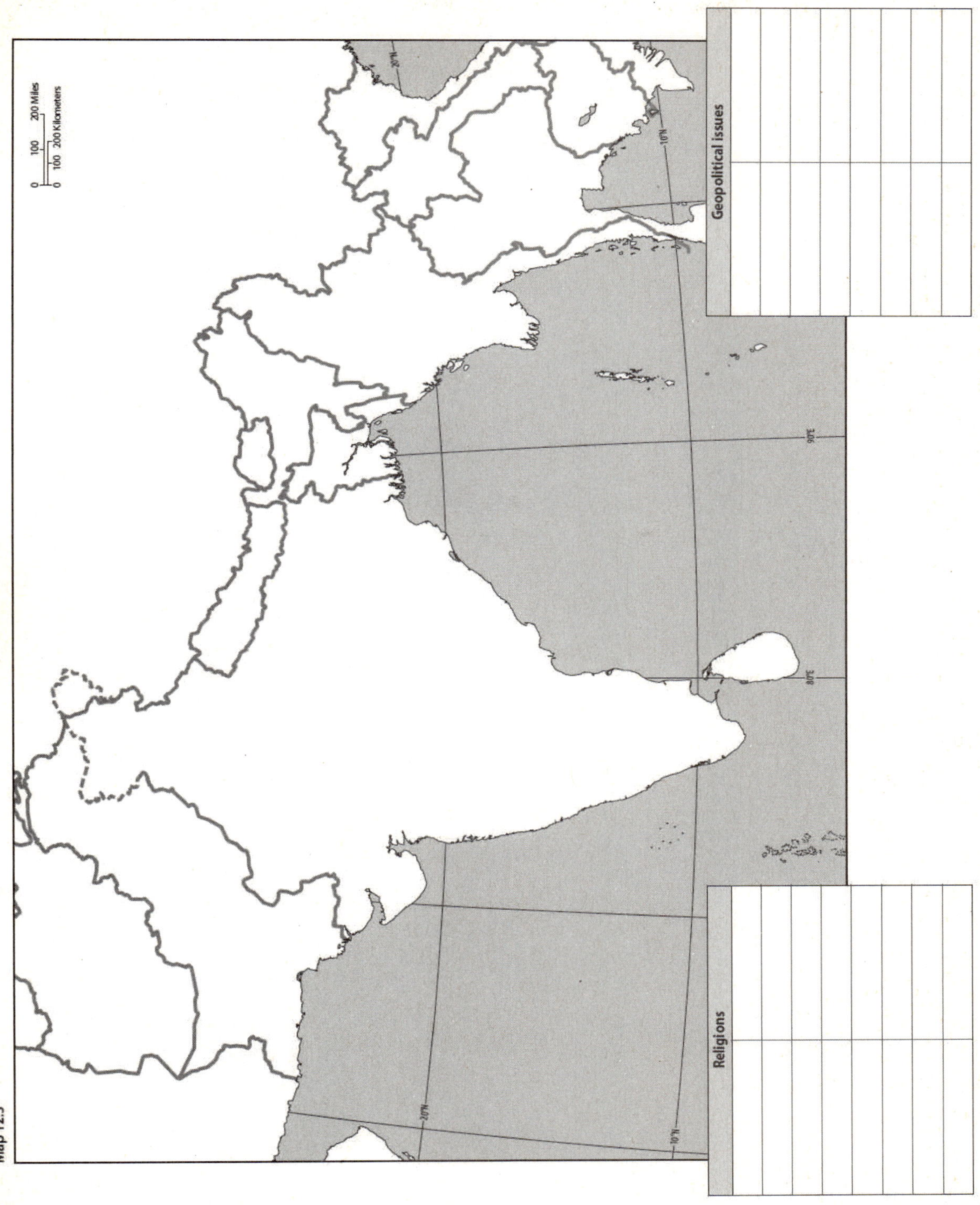
Map 12.3
0 100 200 Miles
0 100 200 Kilometers
20°N
10°N
90°E
80°E
Geopolitical issues
Religions

Exercise Two: South Asian Physical Geography and Population

Using Figure 12.1 South Asia (physical landscape) (p. 345), Figure 12.8 Population Map of South Asia (p. 350), and Mapping Workbook Map 12.4, complete the following exercise.

Using Figure 12.1 "South Asia" (physical landscape) (p. 345) as a reference, use a colored pencil to shade in the low elevation regions of South Asia (0–500 feet above sea level). Using a different colored pencil, shade in the higher elevation regions of South Asia (500–2000 feet above sea level). Using a pencil that is a different color from the first two you used, shade in the highest elevation regions of South Asia (4000 + feet above sea level). Mark the shading scheme in the map's legend. Now, label the major mountain ranges, plains, plateaus and deserts. Use a different colored pencil to draw in and label the location of the major South Asian rivers. Mark the symbol for rivers in the map's legend. Once you have done this, answer the following questions.

1. Describe the location of the low elevation regions in South Asia.

____

____

____

____

2. Describe the location of the higher elevation regions in South Asia.

____

____

____

____

3. Describe the location of the highest elevation regions in South Asia.

____

____

____

____

4. How does the location of rivers correlate with the high and low elevations within the region?

____

____

____

____

Using Figure 12.8 "Population Map of South Asia" (p. 350) as a reference, use a red or black (or a color that will contrast with the elevation shading pattern you established) colored pencil to pattern the locations of South Asian population settlement. Mark the patterning scheme in the map's legend. Once you have done this, answer the following questions.

5. Where are the South Asian regions that possess the highest population densities?

____

____

6. What do you think accounts for the clustering of people into these regions?

7. Compare the patterns of elevation and population settlement on your map. Do physical features seem to play a role in high population densities? If so, what are the physical characteristics of the landscape that appear to promote population?

8. Where are the Asian regions that possess the lowest population densities?

9. What do you think accounts for the clustering of people in these regions?

10. Do physical features seem to play a role in low population densities? If so, what are the physical characteristics of the landscape that appear to limit population settlement?

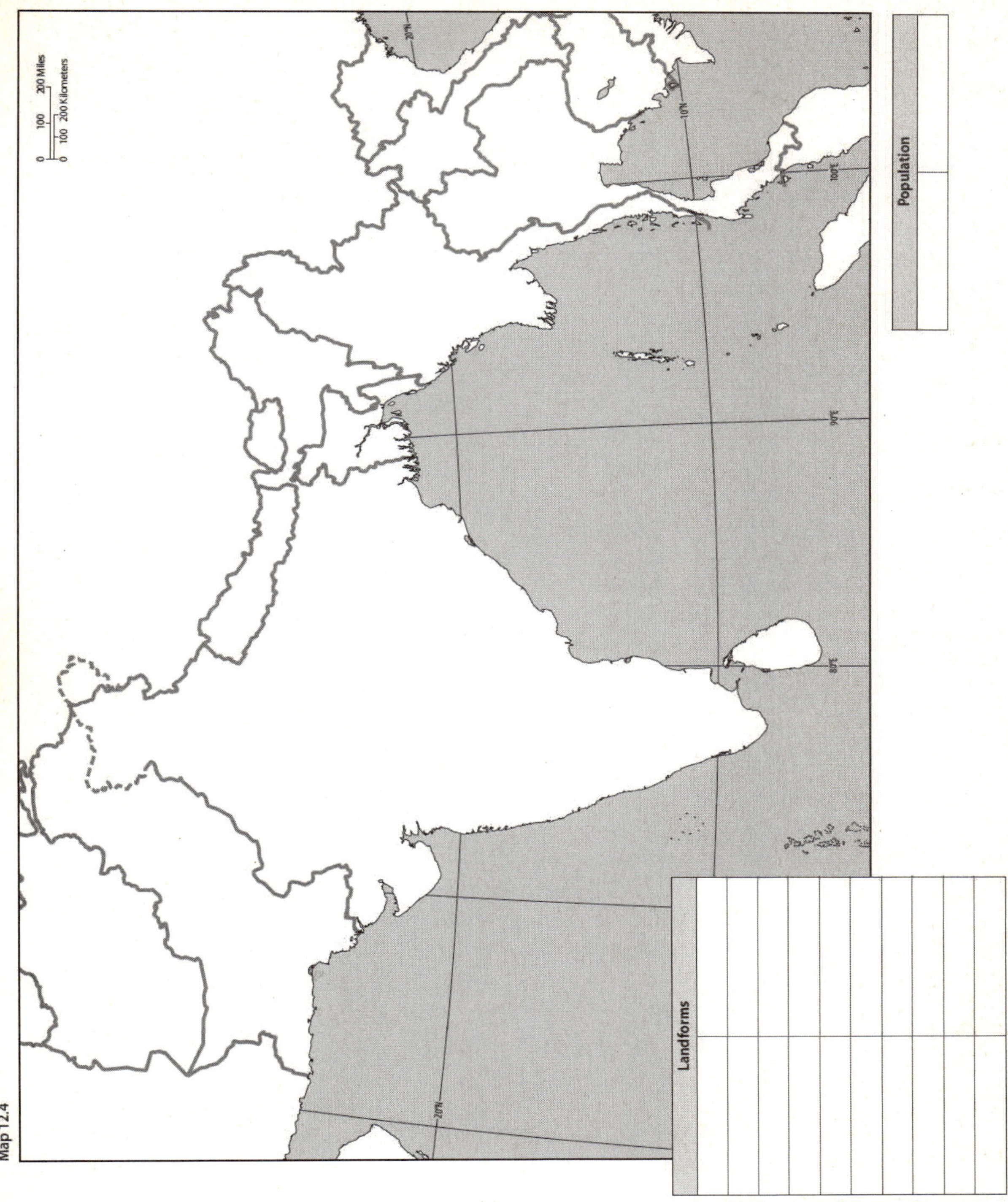
Map 12.4
0 100 200 Miles
0 100 200 Kilometers
Population
Landforms

Chapter Thirteen: Southeast Asia Mapping Workbook Exercises

Identify the following features on workbook Maps 13.1 and 13.2

Identify and label the following countries on Map 13.1

Brunei
Burma (Myanmar)
Cambodia
Indonesia
Laos
Malaysia
Papua New Guinea
Philippines
Sabah (state)
Sarawak (state)
Singapore
Thailand
East Timor
Vietnam

Identify and label the following cities on Map 13.1

Ambon
Bandar Seri Begawan
Bandung
Bangkok
Banjarmasin
Cebu
Da Nang
Davao
Dili
Haiphong
Hanoi
Ho Chi Min City
Ipoh
Jakarta
Kuala Lumpur
Kuching
Makassar
Mandalay
Manila
Medan
Palembang
Phnom Penh
Quezon City
Semarang
Singapore
Songkhia
Sorong
Surabaya
Vientiane
Yangon

Identify and label the following physical features on Map 13.2

Rivers and Other Bodies of Water

Chao Phraya River
Irrawaddy River
Mekong River
Red River
Salween River
Tonle Sap River
Andaman Sea
Banda Sea
Bay of Bengal
Celebes Sea
Gulf of Thailand
Gulf of Tonkin
Indian Ocean
Java Sea
Philippine Sea
South China Sea

Islands and Island Chains

Lesser Sunda Islands
Paracel Islands
Spratly Islands
Bali
Borneo
Flores
Java
Lombok
Luzon
Madura
Mindanao
Negros
Palawan
Seram
Sulawesi
Sumatra
Sumba
Sumbawa
Timor
Visayas

Mountains, Mountain Ranges, and Miscellaneous Physical Features

Annam Mountains
Arakan Mountains
Barisan Mountains
Dieng Plateau
Iran Mountains
Khorat Plateau
Mt. Kerinci
Mt. Kinabalu
Mt. Pinatubo
Mt. Semeru
Mt. Victoria

Map 13.1

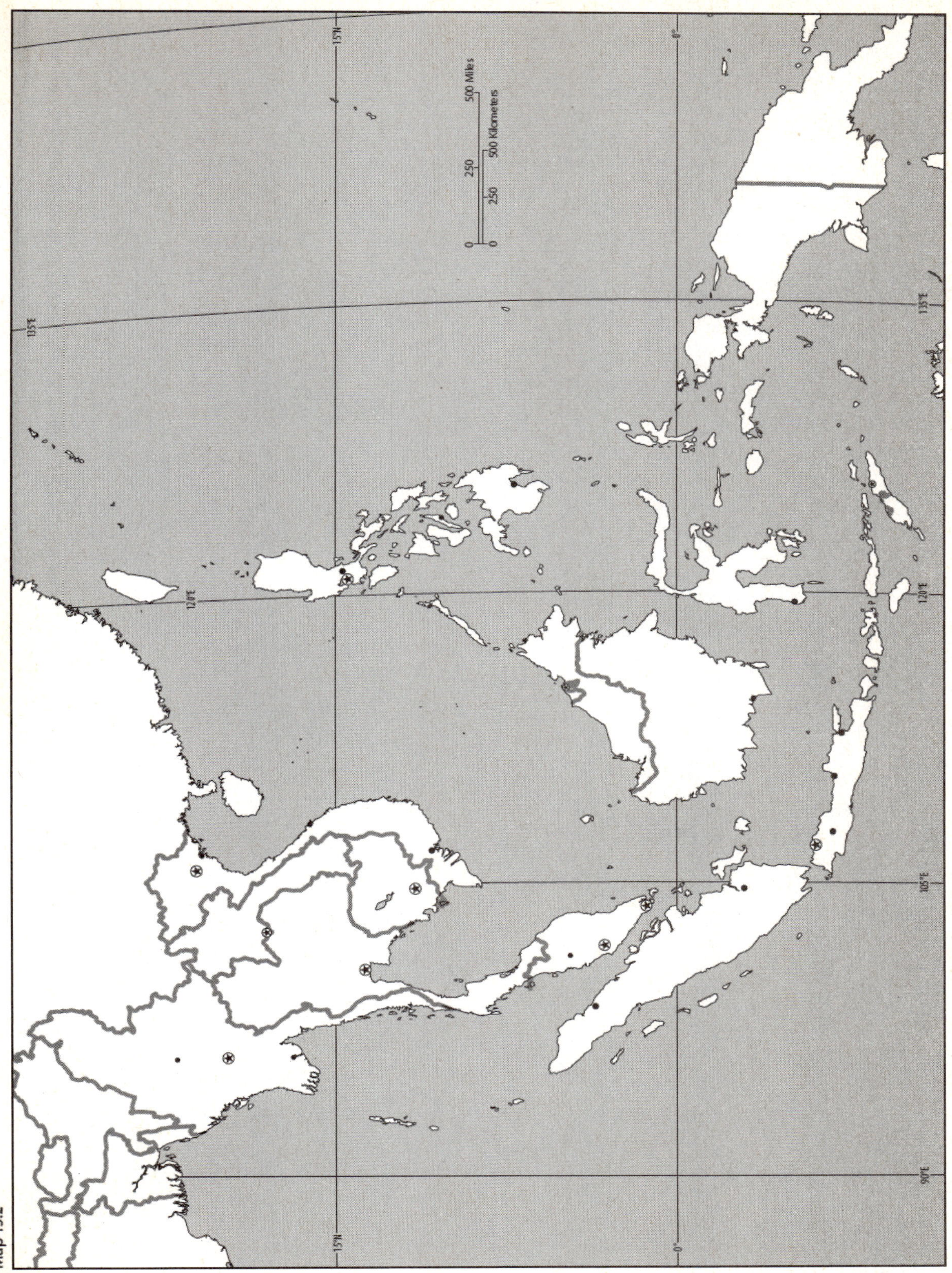

Map 13.2

Exercise One: Southeast Asian Population Settlement and Environmental Issues

Using Figure 13.1 Southeast Asia (physical landscape) (pp. 374-375), Figure 13.11 Population Map of Southeast Asia (p. 382), Figure 13.3 Environmental Issues in Southeast Asia (p. 377), and Mapping Workbook Map 13.3, complete the following exercise.

Using Figure 13.1 "Southeast Asia" (physical landscape) (pp. 374-375) as a reference, use a colored pencil to shade low elevation regions of both Mainland and Insular Southeast Asia (0–500 feet above sea level). Using a different colored pencil, shade in the higher elevation regions (500–2000 feet above sea level). Using a different colored pencil from the first two, shade in the highest elevation regions (4000 + feet above sea level). Mark the shading scheme in the map's legend and label the major mountain ranges. Use a different colored pencil to draw in and label the location of the major Southeast Asian rivers. Once you have done this, answer the following questions.

1. Describe the location of the low elevation regions in Southeast Asia.

____

____

2. Describe the location of the higher elevation regions in Southeast Asia.

____

____

3. Describe the location of the highest elevation regions in Southeast Asia.

____

____

Using Figure 13.11 "Population Map of Southeast Asia" (p. 382) as a reference, use a red or black colored pencil to pattern the locations of South Asian population settlement. Mark the patterning scheme in the map's legend. Once you have done this, answer the following questions.

4. Where are the <u>Mainland</u> Southeast Asian regions that possess the highest population densities?

____

____

5. Compare the patterns of elevation, physical features, and population settlement of <u>Mainland</u> Southeast Asia that you drew on your map. Do the <u>Mainland</u> Southeast Asian physical features seem to play a role in high population densities in this region? If so, what are the physical characteristics of the landscape that appear to promote population settlement?

____

____

____

____

6. Where are the <u>Insular</u> Southeast Asian regions that possess the highest population densities?

____

____

7. Compare the patterns of elevation, physical features, and population settlement of Insular Southeast Asia that you drew on your map. Do the Insular Southeast Asian physical features seem to play a role in high population densities in this region? If so, what are the physical characteristics of the landscape that appear to promote population settlement?

Using Figure 13.3 "Environmental Issues in Southeast Asia" (p. 377) as a reference, use a red or black (or a color that will contrast with the elevation and population shading patterns you established) colored pencil to pattern in the locations of Southeast Asian tropical destruction. Using a different colored pencil, pattern in the locations of the Southeast Asian regions that have experienced the greatest amount of coastal pollution. Mark the patterning scheme in the map's legend. Once you have done this, answer the following questions.

8. Where are the greatest regions of tropical forest destruction in Mainland Southeast Asia?

9. Where are the greatest regions of tropical forest destruction in Insular Southeast Asia?

10. Where are the greatest regions of coastal pollution in Mainland Southeast Asia?

11. Where are the greatest regions of coastal pollution in Insular Southeast Asia?

12. Examine the pattern of population settlement in Mainland Southeast Asia. Is there a correlation between population settlement and tropical forest destruction? If so, how has the Mainland Southeast Asian population affected tropical deforestation in the region?

13. Is there a correlation between population settlement and coastal pollutions? If so, how has the Mainland Southeast Asian population affected coastal pollution in the region?

14. Examine the pattern of population settlement in Insular Southeast Asia. Is there a correlation between population settlement and tropical forest destruction? If so, how has the Insular Southeast Asian population affected tropical deforestation in the region?

---

15. Is there a correlation between population settlement and coastal pollutions? If so, how has the Insular Southeast Asian population affected coastal pollution in the region?

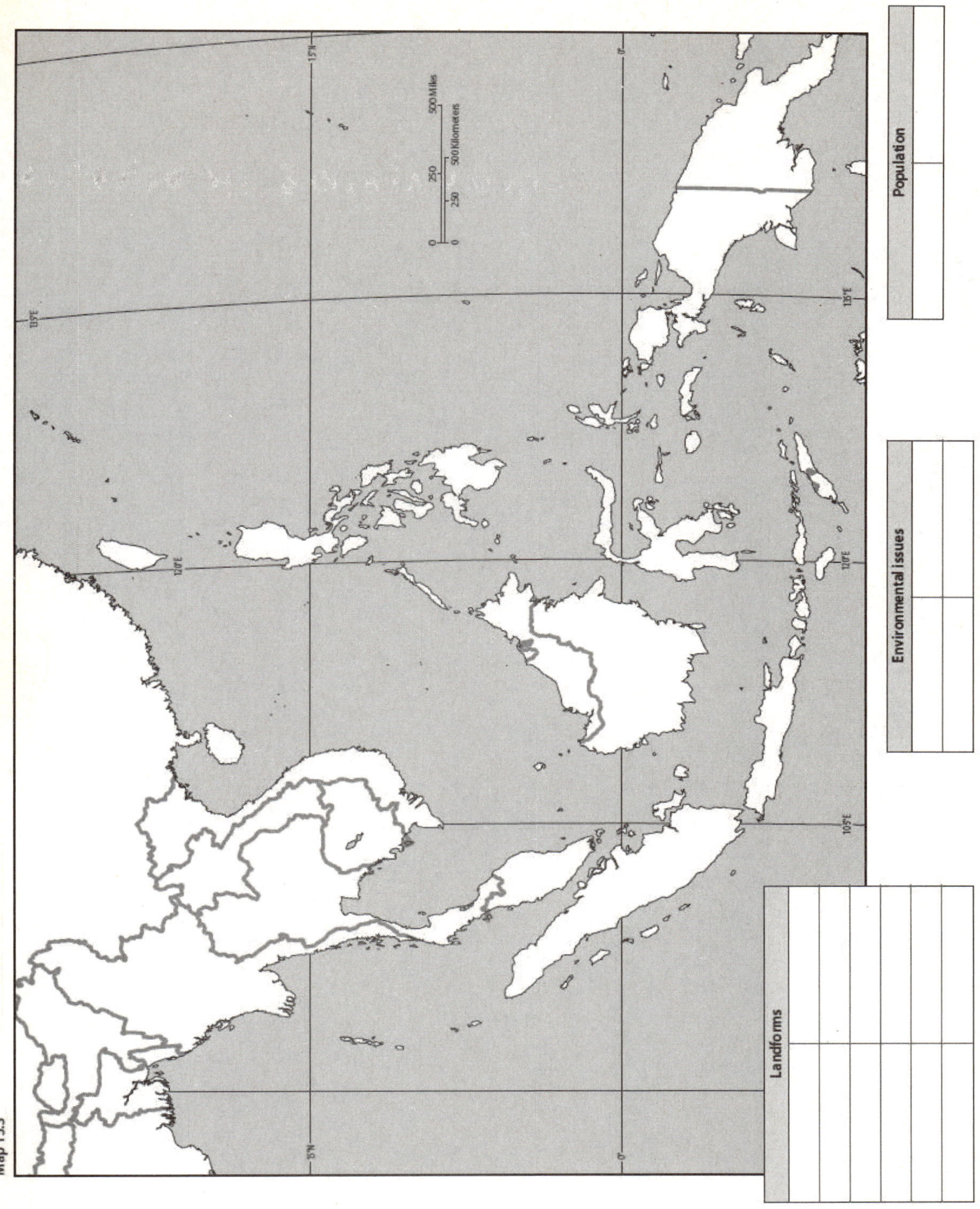
Map 13.3
Population
Environmental Issues
Landforms
0 250 500 Miles
0 250 500 Kilometers
15°N
0°
135°E
120°E
105°E

Exercise Two: Southeast Asian Development Comparison

Using Table 13.2 "Development Indicators" (p. 397) and Mapping Workbook map 13.4, complete the following exercise.

Using the GNI Per Capita, PPP 2007 column in Table 13.2 "Development Indicators" (p. 397) as a reference, use a colored pencil to shade in the Southeast Asian countries possessing an annual per capita GNI of 1,000–2,000. Using a different colored pencil, shade in the Southeast Asian countries possessing an annual per capita GNI of 2,001–3,000. Using a different colored pencil, shade in the Southeast Asian countries possessing an annual per capita GNI of 3,001–4,000. Using a different colored pencil, shade in the Southeast Asian countries possessing an annual per capita GNI of 4,001 and above. Mark your shading scheme in the map's legend.

Using the Human Development Index (HDI), 2008 column from Table 13.2 "Development Indicators" (p. 397) as a reference, use a ruler to draw respective proportional symbols representing the HDI on their respective countries. Have a 1/2 inch (on one side) square represent an HDI of 0.82 or greater, a 1/4 inch (on one side) square represent an HDI of 0.702–0.81, a 1/8 inch (on one side) square represent an HDI of 0.593–0.701, and a 1/16 inch (on one side) square represent an HDI of 0.483–0.592. Use a red or black colored pencil (or a color that will contrast with the shading scheme you created above) to shade in the proportional symbols below.

Study the map you created and answer the following questions.

1. Define GNI.

____

____

2. Which Southeast Asian nations possess the highest annual per capita GNI?

____

____

3. Which Southeast Asian nations possess the lowest annual per capita GNI?

____

____

4. Is there an overall regional pattern of annual per capita GNI for Southeast Asia? If so, what is it?

____

____

5. Using GNI as a development indicator, which Southeast Asian region appears wealthier, Mainland or Insular?

____

6. If one region appears wealthier than the other, what might be the explanation? If the regions are about the same in their levels of development as measured by GNI, what might be the explanation for this similar level of development?

____

____

____

____

7. Now examine the HDI proportional symbols you placed on your map. Define HDI.

____

____

8. Which Southeast Asian nations possess the highest HDI?

9. Which Southeast Asian nations possess the lowest HDI?

10. Is there an overall regional pattern of HDI for Southeast Asia? If so, what is it (e.g., does the region display a sub-region that appears to possess a higher HDI)?

11. Using HDI as a development indicator, which Southeast Asian region appears wealthier, Mainland or Insular?

12. If one region appears wealthier than the other, what might be the explanation? If the regions are about the same in their levels of development as measured by HDI, what might be the explanation for this similar level of development?

13. Compare and contrast the HDI and per capita GNI data you mapped. Is there a correlation (positive or negative) between the HDI and per capita GNI of Southeast Asian nations? If so, what is it?

14. If there was a correlation, what might explain it?

15. If there was no correlation at all, what might explain it?

16. Which of these measurements, HDI or GNI, do you believe to be a better measure of development? Why?

Table

GNI	Color
1,000–2,000	
2,001–3,000	
3,001–4,000	
4,001 and above	

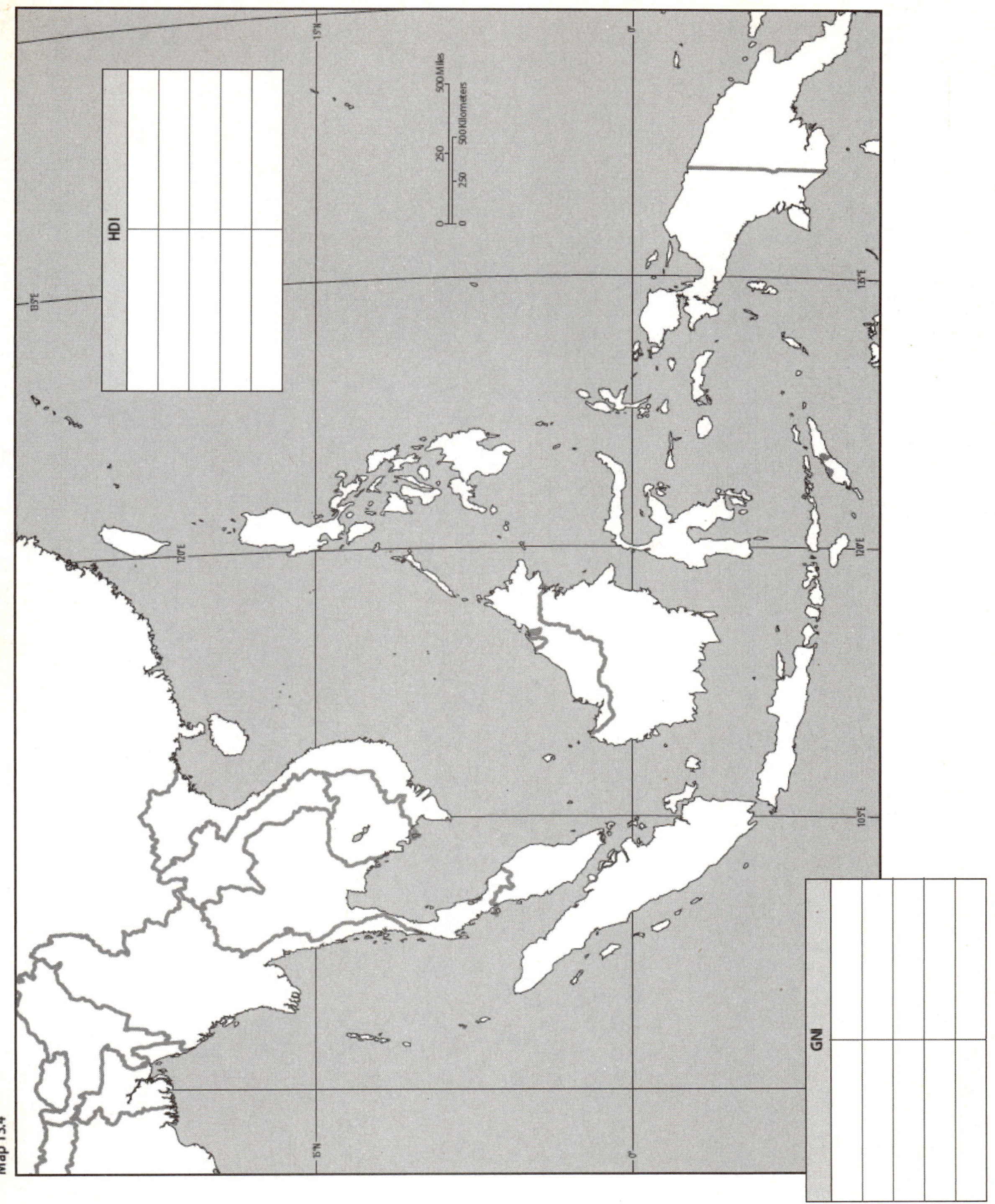
Map 13.4
HDI
GNI
0 250 500 Miles
0 250 500 Kilometers
15°N
0°
135°E
120°E
105°E

Chapter Fourteen: Australia and Oceania Mapping Workbook Exercises

Identify the following features on workbook Maps 14.1 and 14.2

Identify and label the following countries on Map 14.1

American Samoa
Australia
Cook Islands
Federated States of Micronesia
Fiji
French Polynesia
Guam
Hawaiian Islands
Kiribati
Marquesas Islands
Marshall Islands
Nauru
New Caledonia
New Zealand
Niue
Northern Mariana Islands
Palau
Papua New Guinea
Samoa
Society Islands
Solomon Islands
Tahiti
Tokelau
Tonga
Tuvalu
Vanuatu
Wallis and Futuna

Identify and label the following cities on Map 14.1

Adelaide
Auckland
Brisbane
Cairns
Canberra
Christchurch
Funafuti
Hobart
Honiara
Koror
Majuro
Melbourne
Noumea
Nuku'alofa
Pago Pago
Palikir
Papeete
Perth
Port Moresby
Port-Vila
Suva
Sydney
Tarawa
Wellington
Yaren

Identify and label the following physical features on Map 14.2

Arafura Sea
Bass Strait
Bismarck Sea
Cook Strait
Coral Sea
Darling Ranges
Flinders Range
Great Artesian Basin
Great Barrier Reef
Great Dividing Range
Great Sandy Desert
Gulf of Carpentaria
Indian Ocean
Kimberly Plateau
Nullarbor Plain
MacDonnell Range
Melanesia
Micronesia
North Island
Polynesia
Solomon Sea
South Island
Spencer Gulf
Tasman Sea
Tasmania
Torres Strait

Equator
Tropic of Capricorn
Tropic of Cancer
International Date Line
International Date Line
0 500 1,000 Miles
0 500 1,000 Kilometers
15°N
0°
15°S
30°S
45°S
135°W
150°W
165°W
180°
165°E
150°E
135°E
120°E

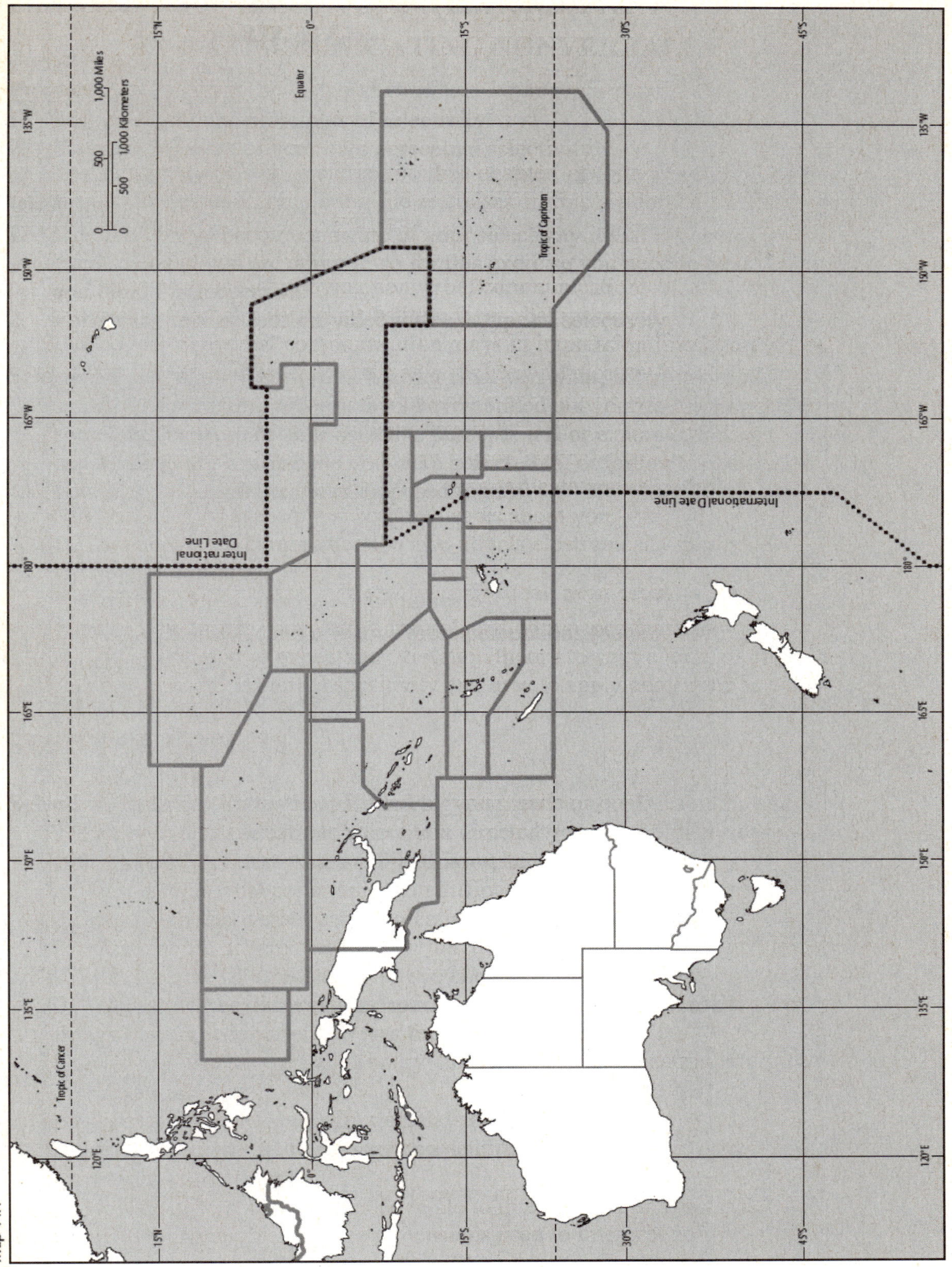

Map 14.1

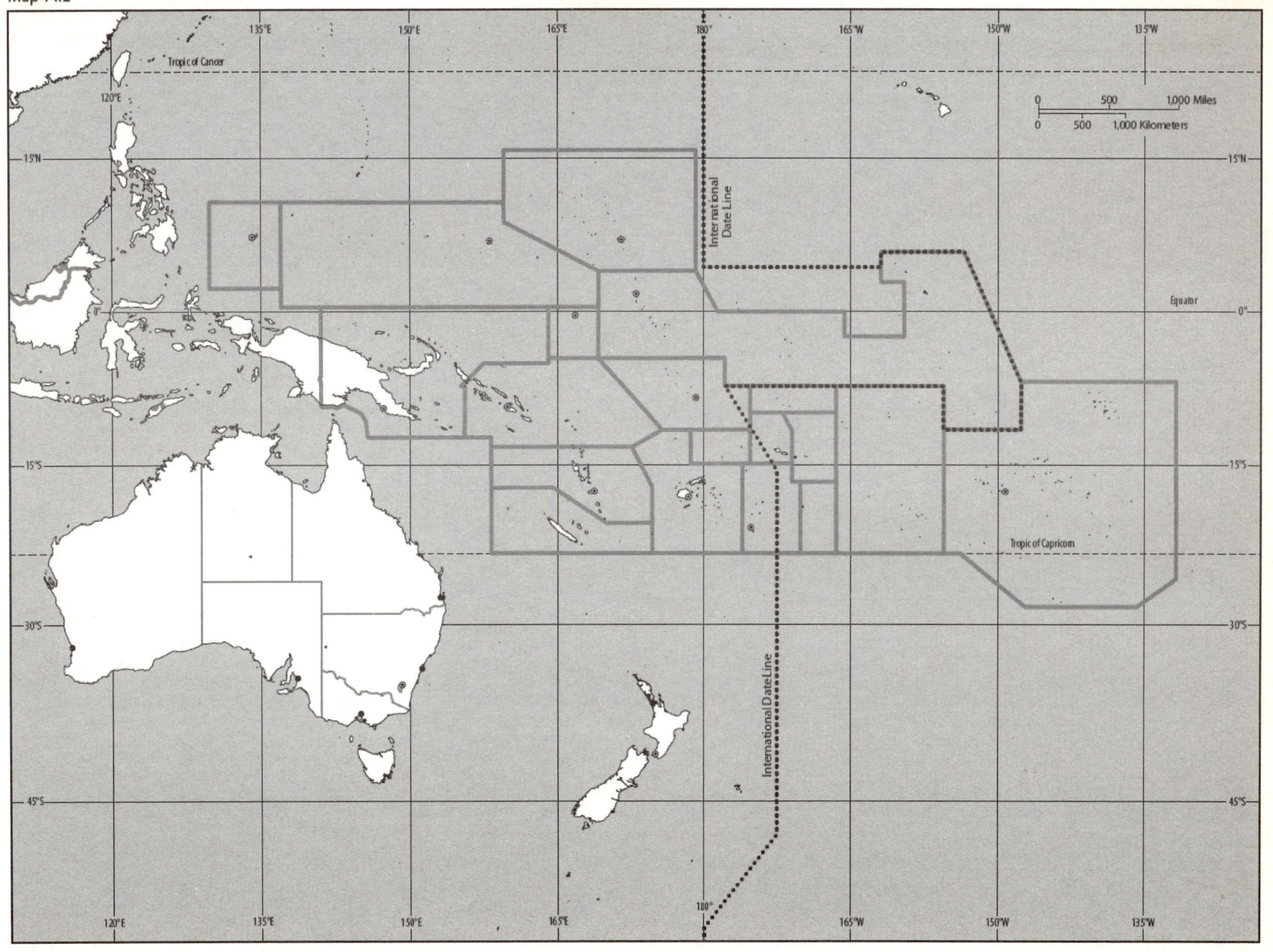
Map 14.2
Tropic of Cancer
Equator
Tropic of Capricorn
International Date Line
International Date Line
0 500 1,000 Miles
0 500 1,000 Kilometers
15°N
0°
15°S
30°S
45°S
120°E
135°E
150°E
165°E
180°
165°W
150°W
135°W

Exercise One: Population and Linguistic Diffusion in Australia and Oceania

Using Figure 14.20 Peopling the Pacific (p. 416), Figure 14.27 Language Map of Australia and Oceania (p. 422), and Mapping Workbook Map 14.3, complete the following exercise.

Using Figure 14.20 "Peopling the Pacific" (p. 416) as a reference, create an isochronic (line of equal time) map. Begin by using a colored pencil and draw a line around the regions of Australia and Oceania that were settled prior to 2500 BCE. Label this line "2500 BCE." Using a different colored pencil, draw a line indicating the regions of Australia and Oceania that were settled by 1200 BCE. Label this line "1200 BCE." Using a different colored pencil, draw a line indicating the regions of Australia and Oceania that were settled by 200 BCE. Label this line "200 BCE." Using a different colored pencil, draw a line indicating the regions of Australia and Oceania that were settled by 400 CE. Label this line "400 CE." Using a different colored pencil, draw a line indicating the regions of Australia and Oceania that were settled by 800 CE. Label this line "800 CE." Using a different colored pencil, draw arrows indicating the general paths of human migration throughout the region. Mark the line and arrow symbols in the map's legend. Once you have done this, answer the following questions.

1. Describe the extent of settlement in Australia and Oceania by 2500 BCE. Specifically, which groups of islands were inhabited by this period?

____

____

____

____

2. Describe the extent of settlement in Australia and Oceania by 1200 BCE. Specifically, which groups of islands were inhabited by this period?

____

____

____

____

3. Describe the extent of settlement in Australia and Oceania by 200 BCE. Specifically, which groups of islands were inhabited by this period?

____

____

____

____

4. Describe the extent of settlement in Australia and Oceania by 400 CE. Specifically, which groups of islands were inhabited by this period?

____

____

____

____

5. Describe the extent of settlement in Australia and Oceania by 800 CE. Specifically, which groups of islands were inhabited by this period?

____

____

____

____

Using a straight edge, draw a line between point "A" and point "B" on the map. Using the map's scale, determine the distance between these two points. Assuming that point "A" represents the starting location of settlement in this region and point "B" represents the general westward limit of expansion by 800 CE, calculate the average rate of population diffusion in miles per century.

6. Between 2500 BCE and 1200 BCE what was the approximate distance that humans had diffused throughout the region?

___

7. What would be the average rate of diffusion in miles per century?

___

8. Between 1200 BCE and 200 BCE what was the approximate distance that humans had diffused throughout the region?

___

9. What would be the average rate of diffusion in miles per century?

___

10. Between 200 BCE and 400 CE what was the approximate distance that humans had diffused throughout the region?

___

11. What would be the average rate of diffusion in miles per century?

___

12. Between 400 CE and 800 CE what was the approximate distance that humans had diffused throughout the region?

___

13. What would be the average rate of diffusion in miles per century?

___

14. Between 2500 BCE and 800 CE what was the approximate distance that humans had diffused throughout the region?

___

15. What would be the average rate of diffusion in miles per century?

___

16. During which of these periods did humans most rapidly expand throughout the region?

___

17. During which of these periods did humans diffuse the most slowly?

___

Using Figure 14.27 "Language Map of Australia and Oceania" (p. 422) as a reference, use a colored pencil to shade in the regions of Papuan-speaking peoples. Using a different colored pencil, shade in the regions of Austronesian-speaking peoples. Using a different colored pencil, shade in the regions of combined Papuan and Austronesian-speaking peoples. Using a different colored pencil, shade in the areas of persisting indigenous languages. Mark your shading scheme on the map's legend. Once you have done this, answer the following questions.

18. Where are the majority of the Papuan-speaking people located within Australia and Oceania?

____

____

19. Where are the majority of the Austronesian-speaking people located within Australia and Oceania?

____

____

20. Where are the majority of the combined Papuan and Austronesian-speaking people located within Australia and Oceania?

____

____

21. Where are the majority of persisting indigenous languages located within Australia and Oceania?

____

____

22. Examine the Australian and Oceanic linguistic patterns and compare them with the isochronic lines of diffusion you drew on your map. Based on the mapped linguistic patterns, can you determine the general area of origin and period of expansion of Papuan speaking peoples throughout the region?

____

____

____

____

23. Based on the mapped linguistic patterns, can you determine the general area of origin and period of expansion of Austronesian speaking peoples throughout the region?

____

____

____

____

24. Based on the mapped linguistic patterns, can you determine the general areas of origin and period when both the Papuan and Austronesian languages became prevalent within the same Oceanic region?

____

____

____

____

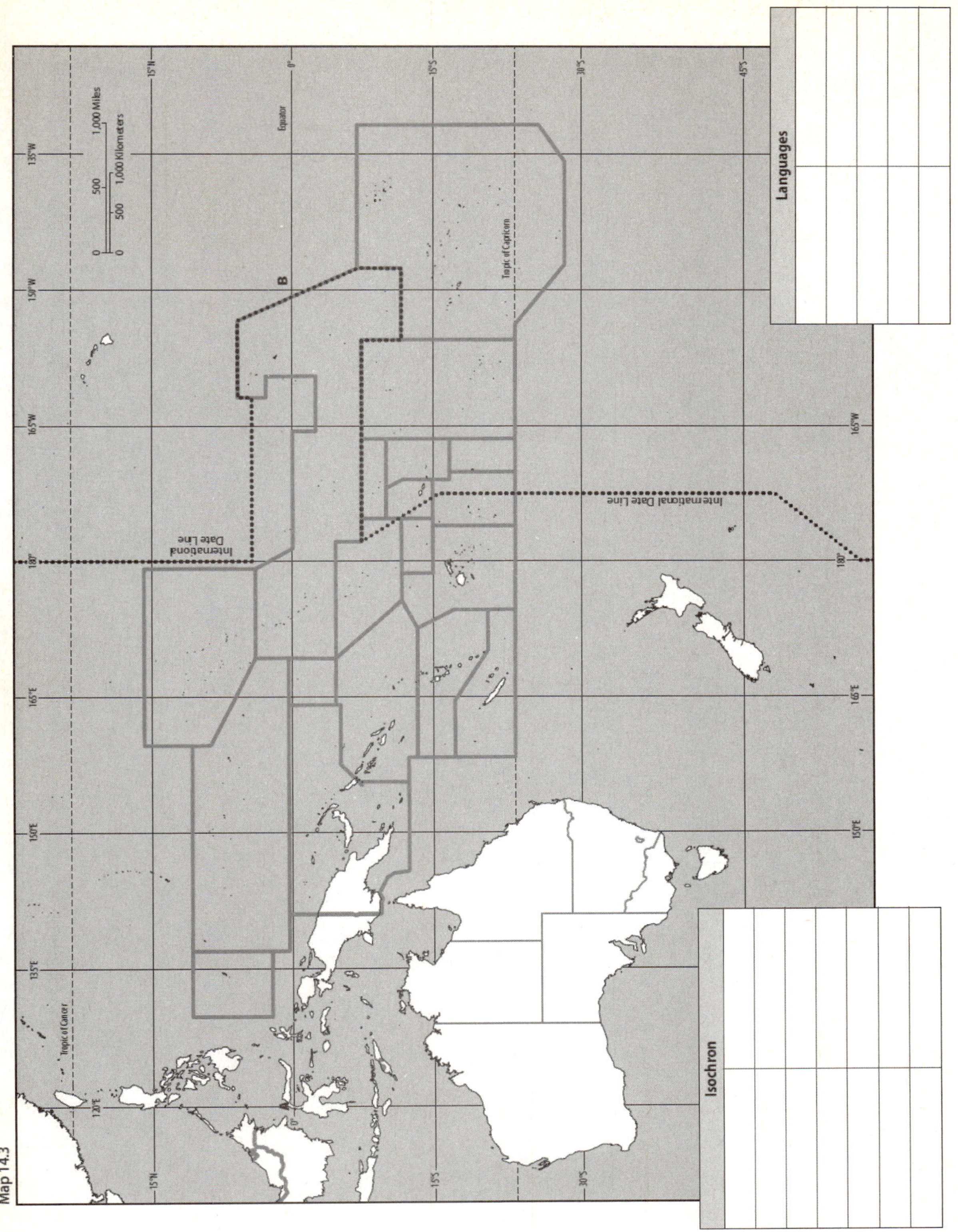
Map 14.3
Languages
Isochron
B
Equator
Tropic of Capricorn
Tropic of Cancer
International Date Line
International Date Line
0 500 1,000 Miles
0 500 1,000 Kilometers
15°N
0°
15°S
30°S
45°S
120°E
135°E
150°E
165°E
180°
165°W
150°W
135°W

Exercise Two: Comparative Patterns of Oceania Development

Using Table 14.2 Development Indicators (p. 429) and Mapping Workbook Map 14.4, complete the following exercise.

Using the GNI Per Capita, PPP 2007 column in Table 14.2 "Development Indicators" (p. 429) as a reference, use a colored pencil to shade in the Oceanic countries possessing an annual per capita GNI of 1,000–2,000. Using a different colored pencil, shade in the Oceanic countries possessing an annual per capita GNI of 2,001–3,000. Using a different colored pencil, shade in the Oceanic countries possessing an annual per capita GNI of 3,001–4,000. Using a different colored pencil, shade in the Oceanic countries possessing an annual per capita GNI of 4,001 and greater. Mark your shading scheme in the map's legend.

Using the data from Table 14.2 "Development Indicators" (p. 429) Human Development Index (HDI), 2008 column, take a ruler and draw respective proportional symbols representing the HDI on their respective countries. Have a 1/2 inch (on one side) square represent an HDI of 0.854 and greater, a 1/4 inch (on one side) square represent an HDI of 0.742–0.853, a 1/8 inch (on one side) square represent an HDI of 0.629–0.741, and a 1/16 inch (on one side) square represent an HDI of 0.516–0.628. Use a red or black colored pencil (or a color that will contrast with the shading scheme you created above) to shade in the proportional symbols below.

Study the map you created and answer the following questions.

1. Define GNI.

____

____

2. Which Oceanic nations possess the highest annual per capita GNI?

____

____

3. Which Oceanic nations possess the lowest annual per capita GNI?

____

____

4. Is there an overall regional pattern of annual per capita GNI for Oceania? If so, what is it?

____

____

5. Using GNI as a development indicator, which Oceanic region appears wealthier: Micronesia, Melanesia, or Polynesia?

____

6. If one region appears wealthier than the other, what might be the explanation? If the regions are about the same in their levels of development as measured by GNI, what might be the explanation for this similar level of development?

____

____

____

____

Now examine the HDI proportional symbols you placed on your map.

7. Define HDI.

____

____

8. Which Oceania nations possess the highest HDI?

____

____

9. Which Oceania nations possess the lowest HDI?

____

____

10. Is there an overall regional pattern of HDI for Oceania? If so, what is it (e.g., does the region display a sub-region that appears to possess a higher HDI)?

____

____

11. Using HDI as a development indicator, which Oceanic region appears wealthier: Micronesia, Melanesia, or Polynesia?

____

12. If one region appears wealthier than the other, what might be the explanation? If the regions are about the same in their levels of development as measured by HDI, what might be the explanation for this similar level of development?

____

____

____

____

13. Compare and contrast the HDI and per capita GNI data you mapped. Is there a correlation (positive or negative) between the HDI and per capita GNI of Oceanic nations? If so, what is it?

____

____

____

____

14. If there was a correlation, what might explain it?

____

____

____

____

15. If there was no correlation at all, what might explain it?

16. Which of these measurements, HDI or GNI, do you believe to be a better measure of development? Why?

Table

GNI	Color
1,000–2,000	
2,001–3,000	
3,001–4,000	
4,001 and greater	

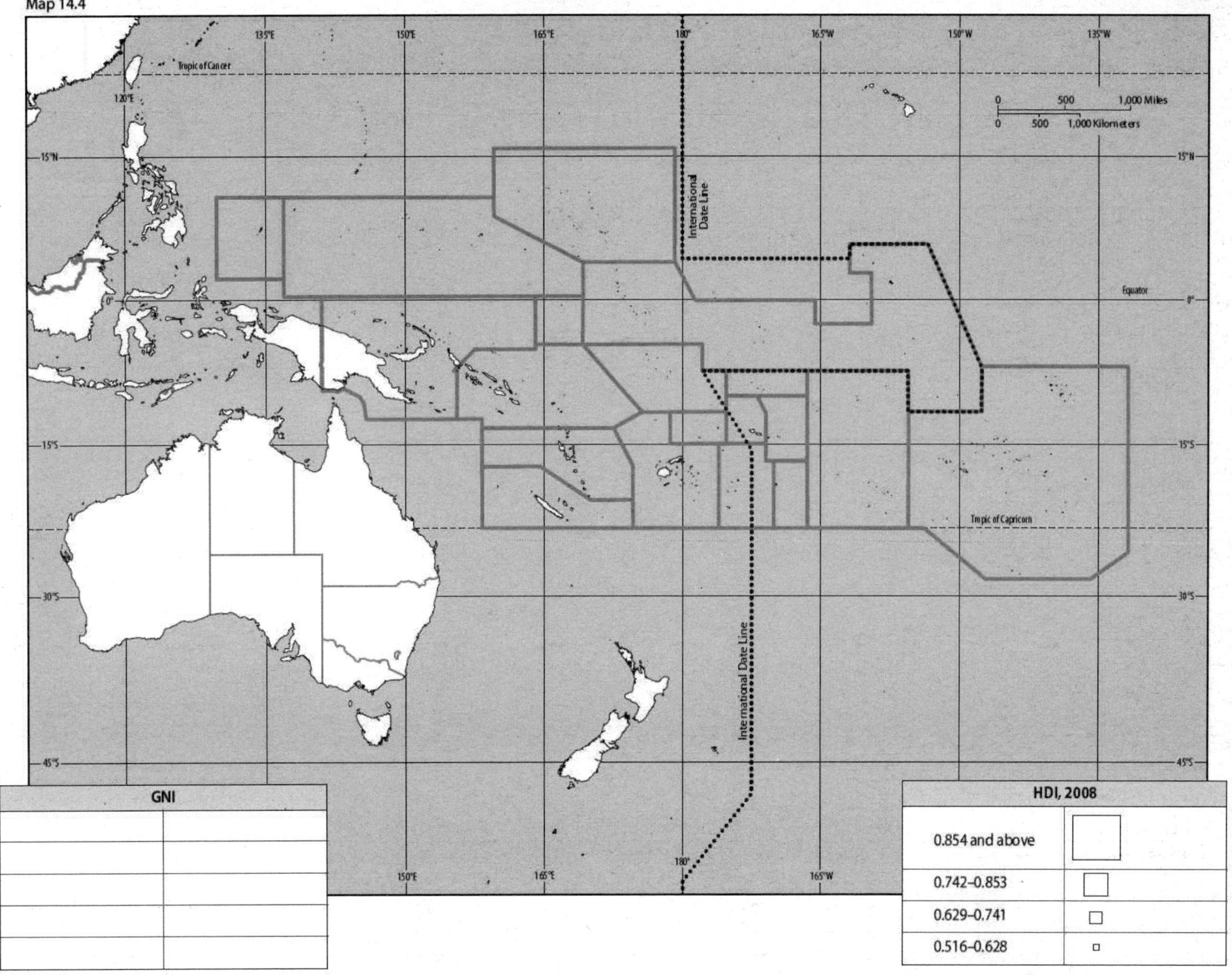
Map 14.4
0 500 1,000 Miles
0 500 1,000 Kilometers
Tropic of Cancer
Equator
Tropic of Capricorn
International Date Line
International Date Line
120°E
135°E
150°E
165°E
180°
165°W
150°W
135°W
15°N
0°
15°S
30°S
45°S
GNI
HDI, 2008
0.854 and above
0.742–0.853
0.629–0.741
0.516–0.628